T0321954

Case-Control Studies

The case-control approach is a powerful method for investigating factors that may explain a particular event. It is extensively used in epidemiology to study disease incidence, one of the best-known examples being the investigation by Bradford Hill and Doll of the possible connection between cigarette smoking and lung cancer. More recently, case-control studies have been increasingly used in other fields, including sociology and econometrics.

With a particular focus on statistical analysis, this book is ideal for applied and theoretical statisticians wanting an up-to-date introduction to the field. It covers the fundamentals of case-control study design and analysis as well as more recent developments, including two-stage studies, case-only studies, and methods for case-control sampling in time. The latter have important applications in large prospective cohorts that require case-control sampling designs to make efficient use of resources. More theoretical background is provided in an appendix for those new to the field.

RUTH H. KEOGH is a Lecturer in the Department of Medical Statistics at the London School of Hygiene and Tropical Medicine.·

D. R. COX is one of the world's pre-eminent statisticians. His work on the proportional hazards regression model is one of the most-cited and most influential papers in modern statistics. In 2010 he won the Copley Medal of the Royal Society 'for his seminal contributions to the theory and application of statistics'. He is currently an Honorary Fellow at Nuffield College, Oxford.

INSTITUTE OF MATHEMATICAL STATISTICS
MONOGRAPHS

IMS Monographs are concise research monographs of high quality on any branch of statistics or probability of sufficient interest to warrant publication as books. Some concern relatively traditional topics in need of up-to-date assessment. Others are on emerging themes. In all cases the objective is to provide a balanced view of the field.

Case-Control Studies

RUTH H. KEOGH
*London School of Hygiene
and Tropical Medicine*

D. R. COX
Nuffield College, Oxford

CAMBRIDGE
UNIVERSITY PRESS

CAMBRIDGE
UNIVERSITY PRESS

University Printing House, Cambridge CB2 8BS, United Kingdom

One Liberty Plaza, 20th Floor, New York, NY 10006, USA

477 Williamstown Road, Port Melbourne, VIC 3207, Australia

314-321, 3rd Floor, Plot 3, Splendor Forum, Jasola District Centre, New Delhi - 110025, India

79 Anson Road, #06-04/06, Singapore 079906

Cambridge University Press is part of the University of Cambridge.

It furthers the University's mission by disseminating knowledge in the pursuit of education, learning and research at the highest international levels of excellence.

www.cambridge.org
Information on this title: www.cambridge.org/9781107019560

First published 2014

A catalogue record for this publication is available from the British Library

Library of Congress Cataloging in Publication data
Keogh, Ruth H., 1979– author.
Case-control studies / Ruth H. Keogh, D.R. Cox.
p. ; cm. – (Institute of Mathematical Statistics monographs ; 4)
Includes bibliographical references and index.
ISBN 978-1-107-01956-0 (hardback)
I. Cox, D.R. (David Roxbee), author. II. Title. III. Series: Institute of
Mathematical Statistics monographs ; 4.
[DNLM: 1. Epidemiologic Methods. WA 950]
RA652
614.4 – dc23   2013047249

ISBN 978-1-107-01956-0 Hardback
ISBN 978-1-108-44267-1 Paperback

Contents

Preface

The retrospective case-control approach provides a powerful method for studying rare events and their dependence on explanatory features. The method is extensively used in epidemiology to study disease incidence, one of the best known and early examples being the investigation by Bradford Hill and Doll of the possible impact of smoking and pollution on lung cancer. More recently the approach has been ever more widely used, by no means only in an epidemiological setting. There have also been various extensions of the method.

A definitive account in an epidemiological context was given by Breslow and Day in 1980 and their book remains a key source with many important insights. Our book is addressed to a somewhat more statistical readership and aims to cover recent developments. There is an emphasis on the analysis of data arising in case-control studies, but we also focus in a number of places on design issues. We have tried to make the book reasonably self-contained; some familiarity with simple statistical methods and theory is assumed, however. Many methods described in the book rely on the use of maximum likelihood estimation, and the extension of this to pseudo-likelihoods is required in the later chapters. We have therefore included an appendix outlining some theoretical details.

There is an enormous statistical literature on case-control studies. Some of the most important fundamental work appeared in the late 1970s, while the later 1980s and the 1990s saw the establishment of methods for case-control sampling in time. The latter have important applications in large prospective cohorts which collect large amounts of information, for example biological samples, but which require case-control sampling designs to make efficient use of resources. There continue to appear in the literature many innovations in case-control study design and, in particular, methodology covering a wide range of areas.

We hope that the book will be useful both to applied and to theoretical statisticians wanting an introduction to the field. Parts of it might be useful

as a basis for a short course for postgraduate students, although we have
not written with that use specifically in mind.

The EPIC-Norfolk study provided some examples; we are grateful for
permission to use this data and thank the staff and participants of the study.

We are grateful for very helpful comments on an initial draft from
the following colleagues and friends: Amy Berrington de Gonzalez, Vern
Farewell, Chris Keogh and Ian White. We are very grateful also to Di-
ana Gillooly at Cambridge University Press for her encouragement and
helpful advice and to Susan Parkinson for her meticulous and constructive
copy-editing.

Ruth Keogh and David Cox

Preamble

This book is about the planning and analysis of a special kind of investigation: a *case-control study*. We use this term to cover a number of different designs. In the simplest form individuals with an outcome of interest, possibly rare, are observed and information about past experience is obtained. In addition corresponding data are obtained on suitable controls in the hope of explaining what influences the outcome. In this book we are largely concerned with binary outcomes, for example indicating disease diagnosis or death. Such studies are reasonably called *retrospective* as contrasted with *prospective* studies, in which one records explanatory features and then waits to see what outcome arises. In retrospective studies we are studying the causes of effects and in prospective studies we are studying the effects of causes. We also discuss some extensions of case-control studies to incorporate temporality, which may be more appropriately viewed as a form of prospective study. The key aspect of all these designs is that they involve a sample of the underlying population that motivates the study, in which individuals with certain outcomes are strongly over-represented.

While we shall concentrate on the many special issues raised by such studies, we begin with a brief survey of the general themes of statistical design and analysis. We use a terminology deriving in part from epidemiological applications although the ideas are of much broader relevance.

We start the general discussion by considering a population of study individuals, patients, say, assumed to be statistically independent. The primary object is to understand the effect of exposures (or treatments or conditions) on an outcome or response. Exposures are represented by a random variable X and the outcome by a random variable Y, where typically X and sometimes Y are vectors. Our interest is in the effect of X on Y. This relationship can be represented in a diagram as in Figure 1. The arrow points from one variable to another, which is in some sense its outcome; that is, the arrow indicates that X has some effect or influence on Y. The direction of the arrow is usually that of time. Such diagrams are described as path diagrams

1

$$X \longrightarrow Y$$

Figure 1 The effect of an exposure X on an outcome Y. The arrow represents statistical dependence directed from X to Y.

or, in the special case where all the variables included are one dimensional, as *directed acyclic graphs* or DAGS. We use path diagrams to help illustrate some issues of study design and analysis.

We define a third class of variables, referred to as *intrinsic variables* and represented by the vector random variable W. Depending on the study setting these variables may affect X or Y or both, an aspect which we discuss further below. The inter-relationships between W, X and Y in a given study setting affect how we estimate the effect of X on Y; that is, the possible involvement of W should be considered. In this specification both X and W are explanatory variables. The distinction between them is one of subject matter and is context dependent and not to be settled by a formal statistical test. For a variable to be considered as an exposure it has to be relevant, even if not realizable, to ask: how would the outcome of an individual have changed had their exposure been different from what it is, *other things being equal*? By contrast, W represents properties of individuals that are immutable in the context in question. We take 'other things being equal' to mean that the intrinsic variable, W, is fixed when one is studying the possible effect on Y of changing the exposure X. Because the variables W are intrinsic, they typically refer to a time point prior to the exposure X.

There are four broad ways in which the systems described above may be investigated:

- by **randomized experiment**;
- by **prospective observational study**, that is, cohort study;
- by **retrospective observational study**, that is, case-control study;
- by **cross-sectional observational study**.

We describe each type of study in turn. In a **randomized experiment** the level of the exposure X for each study individual is assigned by the investigator using a randomizing device, which ensures that each individual is equally likely to receive each of a set of exposure levels. Examples of exposures are medical treatments received by a patient and fertilizer treatments in an agricultural trial. The outcome Y is recorded after a suitable time. The relationship between X, W and Y in a simple randomized experiment is illustrated in Figure 2, where R denotes the randomization process.

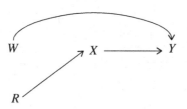

Figure 2 Simple randomized experiment. Randomization R disconnects X and all intrinsic variables W.

In the formulation represented in Figure 2 the exposure X is determined entirely by the randomization process and therefore the exposure for any specific individual is independent of all the intrinsic variables W. As a result, ignoring W produces no systematic distortion when estimating the effect of X on Y.

If W has an effect on Y and the latter is a continuous variable analysed by normal-theory methods then a component of the variance of Y may be in effect eliminated by regression on W and the precision of the resulting assessment of the effect of X thereby enhanced. With a binary Y, the situation with which we are mainly concerned, things are more complicated. In both cases the possibility of an X-by-W interaction may, however, be important.

In an observational study of a population the exposure X is determined by the individual, or as a result of their circumstances, as opposed to being set by the investigator through experimental design. Examples of such exposures are industrial or environmental hazards in the workplace, smoking by individuals and so on. For a patient in a clinical study, W may include gender and age at initial diagnosis or the age at entry into the study. In relatively simple cases X is one dimensional, whereas W is typically multidimensional.

The structure of the data in a **prospective observational study** may appear essentially the same as in a randomized experiment, but in the former there is the crucial distinction that the exposure, X, of each individual is outside the investigator's control. The intrinsic variables W may thus influence X as well as Y.

Suppose that the response variable Y is binary with one outcome having a very low frequency. An individual having this outcome, the outcome of particular interest, is known as a *case*. Then a prospective observational study may be very inefficient, in that large amounts of data may be collected on non-cases, or *controls*, when effectively the same precision for

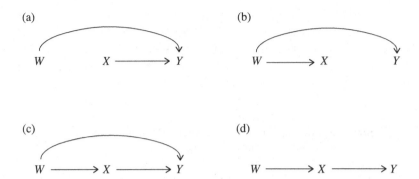

Figure 3 General specification: relationships between exposures X, intrinsic variables W and outcome Y in observational studies.

the case-versus-control comparison would be achieved with many fewer controls. This is one motivation for the third kind of study, the **retrospective observational study**. In this, cases are identified first and then control individuals selected and the variables X and W found retrospectively. The method of selecting controls is crucial and is discussed in more detail in the next chapter. In a retrospective study the object remains, as in a prospective study, to determine the conditional dependence of Y on X given W in the population, but, because of the special methods of selecting individuals used for a retrospective study, this relation is addressed indirectly.

Figure 3 illustrates four different types of interrelationship between X, W and Y which may arise in prospective and retrospective observational studies.

In Figure 3(a) W affects Y but not X. Hence W does not interfere in the effect of X on Y. For a continuous outcome, ignoring W causes no systematic distortion in the estimated effect of X on Y, though controlling for W in a regression adjustment may produce some improvement in precision as noted for randomized experiments. For a binary outcome, however, the situation is different. When the association between Y and X given W is described by the conditional odds ratio between Y and X given W, which is the commonly used effect measure used in case-control studies, this is not the same as the marginal odds ratio between Y and X even if W is not directly connected to X. This is a feature of odds ratios called non-collapsibility and is discussed further in later chapters.

In Figure 3(b) W affects both Y and X, but there is no effect of X on Y given W, indicated by the absence of an arrow from X to Y. An analysis that ignores W would tend to find a spurious non-zero effect of X on Y because of the common effect of W.

Figure 3(c) is the same as Figure 3(b) except for the inclusion of an arrow from X to Y. Failure to account for W in this situation would result in a biased estimate of the effect of X on Y given W. The systematic distortion when there are two paths between X and Y as in Figure 3(c) is called *confounding*. By conditioning on W, which is necessary to assess the effect of changing X while leaving the past unchanged, we close the extra path from X to Y via W and obtain an unbiased estimate of the effect of X on Y.

In Figure 3(d) W affects X but not Y. That is, Y is independent of W given X. Then aspects of the conditional relation between Y and X given W are correctly estimated by ignoring W; that is, it is not necessary to account for W in the analysis, except possibly to examine X-by-W interactions.

Another approach to investigation is by a **cross-sectional observational study**, in which data are collected that refer only to the status of individuals at the instant of observation. This may give useful information on correlations but, in such a study, if two variables are correlated then it is in principle impossible to say from the study alone which variable is the response and which is explanatory; any conclusion of this sort must rest on external information or assumptions. We do not consider cross-sectional studies further here.

In observational studies, an important role of the intrinsic variables W is to adjust for the dependences of X and Y on W, that is, to remove systematic error or *confounding*. This is in contrast with randomized experiments, where the roles of W are precision improvement and interaction detection, always assuming that the randomization has been effective. In some contexts W might be extended to include variables that are not intrinsic features; these variables are defined conceptually prior to the exposure, and, unless accounted for in some way, could distort the estimated association between exposure and outcome. For example, consider a study of the association between a patient treatment and a medical outcome. It might be important to include in W information about other patient medication, use of which may be associated with whether the patient received the treatment of interest and also with the outcome.

A major cause for concern in observational studies is that some components of W may be unobserved. Let W_O and W_U denote the observed and unobserved components respectively. Figure 4 illustrates a situation where both W_O and W_U have arrows to X and Y, that is they are both confounders of the effect of X on Y. Confounding by the observed intrinsic

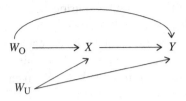

Figure 4 General specification: confounding by observed variables W_O and unmeasured variables W_U.

(a)   (b)

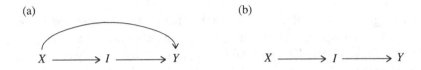

Figure 5 General specification: the presence of an intermediate variable I.

variables W_O can be controlled by conditioning on the W_O. However, because W_U is unobserved its effect cannot be controlled for and the possible role of an unobserved W_U often limits the security of the interpretation. If the arrow from W_U to Y is missing then there is no confounding by the unmeasured variables and conditioning on W_U is not necessary to obtain unbiased estimates of the effect of X on Y. If the arrow from W_U to X is missing but there remains an arrow from W_U to Y then there is no confounding but we may still face a systematic change in the observed association if W_U is ignored in some special situations, notably when the association between X and Y is measured using an odds ratio. This relates to the discussion around Figure 3(a).

We have focused on the role of the intrinsic variables W in the effect of X on Y. We now consider whether the effect of X on Y may possibly act though an intermediate variable, I say (Figure 5). First, in the primary analysis of the effect of X on Y the intermediate response I is ignored, that is, marginalized. In a subsidiary analysis we may consider the mediating effect of I, in particular whether any effect of X on Y is explained largely or even entirely through the effect of X on I, in which case Y is conditionally independent of X given I (Figure 5(b)). In other situations I is itself of interest and is analysed as a response on its own, ignoring Y.

In summary, it is crucial in considering the relation between exposure and outcome to include appropriate conditioning variables W, and to exclude inappropriate ones, and to exclude intermediate variables I when studying the total effect of X on Y.

The ideas sketched here can be developed in many directions, for example to mixtures of study types and to complications arising when there are different exposure variables at different time points for the same individual.

Notes

See Wright (1921) for an early discussion of correlation and causation. Greenland *et al.* (1999) give an introductory account of directed acyclic graphs with an emphasis on epidemiological applications. For discussions of the use of directed acyclic graphs in case-control studies, see Hernán *et al.* (2004), Didelez *et al.* (2010) and Mansournia *et al.* (2013).

1

Introduction to case-control studies

- A case-control study is a retrospective observational study and is an alternative to a prospective observational study. Cases are identified in an underlying population and a comparable control group is sampled.
- In the standard design exposure information is obtained retrospectively, though this is not necessarily the case if the case-control sample is nested within a prospective cohort.
- Prospective studies are not cost effective for rare outcomes. By contrast, in a case-control study the ratio of cases and controls is higher than in the underlying population in order to make more efficient use of resources.
- There are two main types of case-control design; matched and unmatched.
- The odds ratio is the most commonly used measure of association between exposure and outcome in a case-control study.
- Important extensions to the standard case-control design include the explicit incorporation of time into the choice of controls and into the analysis.

1.1 Defining a case-control study

Consider a population of interest, for example the general population of the UK, perhaps restricted by gender or age group. We may call a representation of the process by which *exposures* X and *outcomes* Y occur in the presence of intrinsic features W the *population model*. As noted in the Preamble, such a system may be investigated prospectively or retrospectively; see Figure 1.1. In a prospective or cohort study a suitable sample of individuals is chosen to represent the population of interest, values of (W, X) are determined and the individuals are followed through time until the outcome Y can be observed. By contrast, in a retrospective case-control study, the primary subject of this book, we start with individuals observed to have a specific outcome, say $Y = 1$, whom we call *cases*, and then choose a suitable number of *controls*, often one control for each case. For the

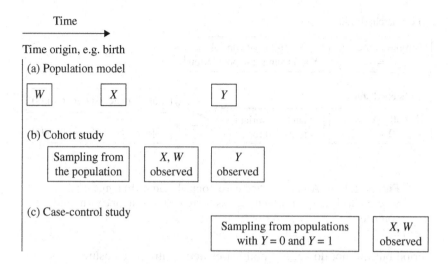

Figure 1.1 (a) Population model with intrinsic features W present; (b) cohort study (prospective); (c) case-control study (retrospective).

controls, members of the population at risk who are not cases, we define Y to be zero. Values of (W, X) are then determined retrospectively on the chosen individuals. The cases are chosen to represent those occurring in the population of interest, and the controls are chosen to represent the part of the population of interest with $Y = 0$.

The essence of a case-control study is that we start with the outcome and look back to find the exposures, that is, the explanatory features of concern. Another core characteristic is that the ratio of cases and controls is not the same as in the population. Indeed typically in the population that ratio is very small because the outcome is rare, whereas in the case-control study it can be as high as one to one. Case-control studies can, however, also be used in situations where the outcome is common rather than rare.

Different procedures for sampling cases and controls lead to several different forms of case-control design. The 'standard' case-control design is as follows and is illustrated in Figure 1.1(c). The cases consist in principle of all individuals in a specific population who have experienced the outcome in question within a specified, usually short, period of time, be that calendar time or age. More generally one might take a random sample of such cases. The control group is sampled from those individuals in the same population who were eligible to experience the outcome during the specified time

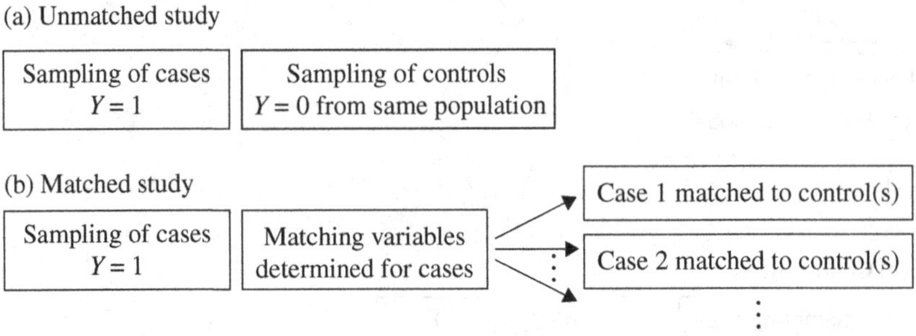

(a) Unmatched study

Sampling of cases $Y = 1$	Sampling of controls $Y = 0$ from same population

(b) Matched study

Figure 1.2 (a) An unmatched case-control study; (b) a matched case-control study, in which the cases are matched to one or more controls.

period but did not do so. For both cases and controls, exposure measures and intrinsic variables are then determined.

Within this standard framework there are two main forms of case-control study, unmatched studies and matched studies; see Figure 1.2. In the first, an *unmatched case-control study*, a shared control group for all cases is selected essentially at random, or, more often, at random given a set of intrinsic variables, perhaps such that the distribution of certain intrinsic variables is similar among the case group and the control group. In the second form, a *matched case-control study*, controls are selected case by case in such a way that they are constrained to match individual cases in certain specified respects, that is, so that to each case is attached one or more controls. Matching on variables that confound the effect of exposure on outcome is a way of conditioning on the confounders. Methods for dealing with confounding in both unmatched and matched studies are discussed in more detail in Section 1.3.

The path diagrams used in the Preamble refer to the study population underlying a case-control study, but they can be extended to incorporate case-control sampling. We define on the underlying study population a binary indicator variable D taking the value 1 for individuals in the case-control sample and taking the value 0 otherwise. Figure 1.3(a) extends previous diagrams to introduce D; we include the possibility of confounding by intrinsic variables W. The arrow from Y to D arises because, in the study population, individuals with $Y = 1$ are much more likely to be sampled to the case-control study than individuals with $Y = 0$. The case-control study corresponds to those with $D = 1$ and hence any analysis is conditional on $D = 1$; the conditioning is indicated by the box around

(a)

(b)

(c)

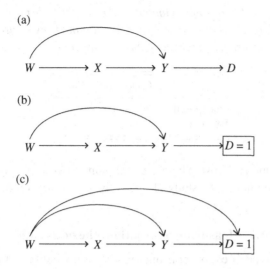

Figure 1.3 (a) Relationships between the variables in an observational study; sampling to the case-control study is denoted by D. (b) Case-control analyses are conditional on $D = 1$. (c) The sampling of individuals to the case-control study may depend on intrinsic variables W.

$D = 1$ in Figure 1.3(b). As noted above, it is common for sampling of controls for a case-control study to depend on intrinsic variables; this is illustrated in Figure 1.3(c) by the arrow from W to D.

Later in this chapter and in the book, we discuss more complex and specialized case-control methods including extensions in which, for example, the case-control study is embedded in a larger investigation or in which the temporal sequence of events is taken into account explicitly.

Case-control studies are often used to study rare outcomes, for which a retrospective study offers clear advantages over a prospective study. A primary advantage is that we do not need to wait many years or even decades to identify cases: for rare events a prospective study requires the follow-up of a large cohort in order to obtain a reasonable number of cases. The assembly of exposure information for a large cohort can be both unnecessarily expensive and time consuming, depending on the nature of the exposures of interest. For rare outcomes a prospective study would involve exposure information being obtained for a large number of non-cases, and in broad terms it is rarely cost effective to have by design comparisons of groups of very unequal size. Case-control studies therefore offer advantages

Table 1.1 *Data from an unmatched case-control study involving n_0 controls and n_1 cases with binary exposure and exposure frequencies r_0 and r_1 respectively.*

	Controls	Cases
Unexposed	$n_0 - r_0$	$n_1 - r_1$
Exposed	r_0	r_1

in terms of time and cost. There are still gains to be made by the use of case-control studies in the study of relatively more common outcomes.

1.2 Measuring association: the odds ratio

The simplest type of exposure is binary, so that any individual is either exposed or unexposed. Examples are whether a person is currently a smoker or non-smoker, and whether blood pressure is above or below a threshold level of interest. In a case-control study with a binary exposure the simplest analysis involves calculating the prevalence of exposure in the case group and in the control group and then examining whether the exposure prevalence differs by case and control status. However, usually it is desirable to obtain a more formal measure of association and certainly something more complex becomes necessary in order to take into account additional variables W, estimate the effects of continuous exposures or, more generally, estimate the effects of exposures of various types while conditioning on potentially a large number of other variables. In this section we introduce the *odds ratio*, which has become the dominant measure of association in case-control studies in the design framework discussed so far.

The data from an *unmatched* case-control study with a binary exposure can be summarized in a 2×2 table as in Table 1.1.

A critical issue of interpretation now arises. The object of study is the dependence of outcome on exposure. But in fact the data have been obtained by choosing individuals with specified outcomes and then observing their exposures.

The odds of a positive exposure among the cases is $r_1/(n_1 - r_1)$ and among the controls is $r_0/(n_0 - r_0)$. Thus the odds ratio is

$$\frac{r_1/(n_1 - r_1)}{r_0/(n_0 - r_0)}; \tag{1.1}$$

an odds ratio of unity indicates no association between exposure and outcome.

Suppose for a moment that the data in Table 1.1 had arisen prospectively rather than retrospectively. If now formally we treat the case-control status as an outcome and the exposure as explanatory, we can calculate the odds of case versus control among exposed individuals as r_1/r_0 and among unexposed individuals as $(n_1 - r_1)/(n_0 - r_0)$. The ratio of these is again (1.1). That is, in a case-control study we may indeed treat the case-control status as an outcome, at least so long as we use the odds ratio as the basis for comparison.

This result is crucial for the analysis of case-control studies in general, including more complex analyses than the above; we discuss it further in Chapter 2. When the outcome is rare in the underlying population an important feature of the odds ratio is that it approximates the ratio of the probability of being a case while exposed and the probability of being a case while unexposed. This ratio, to which the odds ratio is an approximation, is called a *risk ratio* or sometimes a *relative risk*.

The odds ratio in (1.1) refers to the simplest situation, that of a binary exposure, but it extends easily to exposures with more than two categories. The early use of case-control studies focused on binary and categorical exposures; however, the need arose to analyse exposures measured on a continuous scale. Examples include systolic or diastolic blood pressure, usual daily caloric intake and amount of debt. When the exposure is continuous the simple method of analysis outlined above no longer applies unless we divide the range of the continuous exposure into categories using cut points and thus treat it as a categorical variable. To allow for detailed consideration of continuous exposures the *logistic regression* model, which is used to model the probability of an outcome given an exposure in a prospective study, can be extended for use in retrospective case-control studies. This is discussed in detail in Chapter 4.

The odds ratio has become the dominant measure of association in case-control studies in the design framework described so far. Later we discuss a special type of case-control design that enables the direct estimation of *risk ratios*. We also discuss later the introduction of time into case-control studies and show that *rate ratios* can be estimated directly under certain types of case-control sampling.

1.3 Methods for controlling confounding

1.3.1 Preliminaries

The ability to adjust for variables that could confound the association between an exposure and an outcome is an important feature of case-control studies, or indeed of any observational study. The aim is to assess the exposure–outcome association, other things being equal, and so come closer to a causal interpretation. To achieve this we need, so far as is feasible, to condition on features that are explanatory both for the exposure and for the outcomes of concern. Conditioning on confounders is possible at both the design and the analysis stage of a case-control study. The commonly used measure of association is an odds ratio that is conditional on confounding variables.

1.3.2 Unmatched case-control studies

In an unmatched case-control study two simple methods for controlling confounding are the following.

(1) The association between the exposure and the outcome is investigated in *strata* defined by one or more confounders W. For a single binary or categorical exposure and a binary or categorical confounder we can construct tables such as Table 1.1 for each level of the confounder. Odds ratios within each category of the confounding variable can then be compared and, subject to reasonable consistency, combined (see Chapter 2). This approach results in an odds ratio that is conditional on the confounders.

(2) Another possibility is to sample controls in such a way that the distribution of the confounding variables W is the same in the control group as in the case group; this is often referred to as *frequency matching*. Simple comparisons of the exposure prevalence or the mean exposure level within the case and control groups can be made. In this setting, the odds ratio estimated from a table such as Table 1.1 might be referred to as a *marginal* odds ratio because it refers to the marginal association between X and Y in a population with the given distribution of confounders. Note that this is the distribution of confounders in the cases and not in the underlying population of interest. The use of frequency matching goes back to a time before more complex methods for the analysis of case-control data were available, though it remains a popular way of selecting controls. Under frequency matching it will usually be desirable to estimate odds ratios that are conditional on confounders, rather than marginal odds ratios. Provided that there are not too many binary or categorical confounders this can be done by stratified analysis as described above in (1).

It becomes practically difficult to perform either of the above methods when there are many confounders to consider. Furthermore, if there are many confounders, or confounders with multiple levels, then the numbers of cases and controls within strata will become too small for reliable analysis. Neither method accommodates continuous confounders, unless they are

Table 1.2 *Data from a pair-matched case-control study with binary exposure: the number of pairs of each of the four types.*

	Case unexposed	Case exposed
Control unexposed	n_{00}	n_{10}
Control exposed	n_{01}	n_{11}

categorized. It is clear, therefore, that more complex methods for controlling confounding will be needed. The solution is logistic regression, which can be used to adjust for several confounders simultaneously, both categorical and continuous, by regression adjustment. The stratified sampling of controls and the frequency matching of controls based on some confounding variables, so as to over-represent some strata of the confounder, can improve the estimation of conditional odds ratios in logistic regression analyses, and both remain popular methods for sampling controls.

1.3.3 Matched case-control studies

Matching cases to controls on variables that are thought to confound the association between exposure and outcome is a way of directly controlling for confounding at the design stage of a case-control study. The data from a matched case-control study comprise matched sets, each containing a case and one or more controls; matched case-control studies require special methods of analysis.

Consider a pair-matched case-control study, in which each matched set contains one case and one control, and a single binary exposure. The data from this case-control study can be summarized as in Table 1.2. The data in this table can be arranged to obtain the odds ratio n_{10}/n_{01} that is conditional on the confounders used in the matching. The formal justification of this will be outlined in Chapter 3. Case-control pairs in which both the case and the control have the same exposure tell us nothing about the exposure–outcome association; it is only those that are 'discordant' with respect to the exposure that are informative. Table 1.2 can be extended to allow multiple controls per case and categorical exposures, though the calculation to obtain the odds ratios quickly becomes rather more complicated than it is in the pair-matched study with binary exposure.

To allow for continuous exposures, logistic regression can be used to estimate odds ratios in a matched case-control study. The logistic regression model must take account of the matching, and the resulting analysis then becomes a *conditional logistic regression*. Logistic regression for unmatched case-control studies is by contrast described as unconditional.

If there are a large number of potential confounders it may be difficult to obtain a control that matches each case on every factor. It is also important that the main effects of the matching of variables on the outcome cannot be estimated in a conditional analysis of a matched case-control study. Often, therefore, in a matched study, matching is based on a few important and well-established confounders not of independent interest as risk factors for the outcome, and adjustment for additional confounders is postponed to the analysis stage in the conditional logistic regression.

An important use of matching is to control for complex confounders that cannot be quantified for use in a regression adjustment analysis. For example, controls could be matched to cases on area of residence in an attempt to control for a range of environmental and sociological factors. An extreme example of matching is in twin studies, in which the matched case-control sets are pairs of twins, one twin a case and the other acting as control. Another special type of matching occurs when cases are matched to siblings, parents or other close relatives. Family-matched case-control studies are discussed in Chapter 6.

1.4 Temporal aspects

1.4.1 Preliminaries

In this section we discuss how the passage of time is handled in case-control studies. We begin by considering a situation in which we picture the underlying population of interest as in Figure 1.4. The types of case-control study discussed so far involve all, or a random sample of, cases occurring over the time period, which is usually short, and controls drawn from a group who are not cases at the end of that time period. This is sometimes referred to as *cumulative case-control sampling*.

A different possibility in some settings is to compare the cases with a sample of individuals taken at the time origin, that is, the *start* of the period during which cases are observed. This is sometimes referred to as sampling from the study base, and as such has been referred to as *case-base sampling*. Note that the 'control' sample under this approach may contain some individuals who become cases during the time period of interest. Considering a single binary exposure, the data arising from this special case-control sampling scheme can be arranged as in Table 1.3. The odds

Table 1.3 *Data from an unmatched case-control study with binary exposure. The control group is sampled from the study base and may contain some cases. Some individuals may therefore contribute to both the control group and the case group.*

	Controls	Cases
Unexposed	$n_0^* - r_0^*$	$n_1 - r_1$
Exposed	r_0^*	r_1

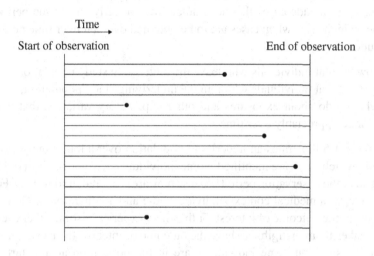

Time

Start of observation   End of observation

Figure 1.4 An underlying population observed over a particular period of time in which some cases arise (•) and all other individuals are observed until the end of the period. The lines indicate the lengths of time for which an individual is in the population.

ratio from this 2 × 2 table is

$$\frac{r_1/(n_1 - r_1)}{r_0^*/(n_0^* - r_0^*)}. \tag{1.2}$$

This *odds ratio* can in fact be shown to be an estimate of the *risk ratio* in the underlying population without any assumption that the case outcome is rare. This is in contrast with the odds ratio estimated using the cumulative sampling approach, as given in (1.1), which estimates the risk ratio only when the cases are rare. This is an example of a special type of case-control study, that we refer to in this book as a *case-subcohort study*. Case-subcohort studies are discussed in detail in Chapter 8, where we elaborate on the interpretation of (1.2) as a risk ratio.

1.4.2 More complex situations

In the situation considered above, illustrated in Figure 1.4, all individuals in the population were in principle observable for the whole of the follow-up period during which cases arose. Sometimes this will be unrealistic, in particular when the time period over which cases are to be ascertained is long. It is for this reason that case-control studies of the types discussed so far usually include cases that accumulate over a fairly short time period. Issues which arise when cases are to be obtained over a longer time period include:

- how to treat individuals who leave the underlying study population;
- how to treat individuals who join the underlying study population;
- what to do about exposures and other explanatory variables that may change appreciably over time.

Figure 1.5 illustrates an underlying population over a long time period, during which cases are identified. Some individuals may not be observed to the end of the observation period because they are censored in some way. For example, in a medical context, individuals die and are therefore no longer at risk for the outcome of interest, or they may experience some other event that makes them ineligible to have the outcome of interest. In an even more complex situation some individuals are in the population at the start of the observation time while others join the population later. Individuals may join the population when they become eligible. For example, eligibility may be restricted by age or geographical region of residence.

A central question is how to select a suitable control group for the cases in these more complex, but not unusual, settings. Suppose that there is no influx into the underlying population, so that all individuals in the population could in principle be observed at the start of the observation period. One (usually unwise) possibility involves using as controls a group of individuals who are observed to survive to the end of the observation period. An odds ratio could technically be defined. However, if the observation period had been shorter then, for cases occurring early in the observation time period, the pool of potential controls would be different from that if the observation period had been longer, because of the changing nature of the

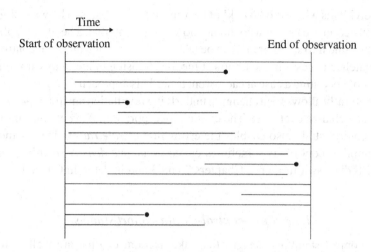

Figure 1.5 An underlying population observed over a particular period of time in which some cases (•) arise and during which individuals can join and leave the population. The lines indicate the lengths for time for which an individual is in the population.

population. In fact, the pool of potential controls for early cases would include future cases. One hopes, therefore, that something better must be possible.

This above approach would also bring difficulties if the exposure changed appreciably over time. Moreover, we would be concerned about whether the comparison with cases would be appropriate; for example, are individuals who survive over a long time period without being censored for some reason intrinsically different from the rest of the underlying population of non-cases? With this approach it is also not clear how to deal easily with influx to the population.

If cases are to be identified over a long time period, it is usually desirable to incorporate 'time' explicitly into the sampling of suitable controls. One solution to the difficulties posed by longer observation periods is to split the long time period up into smaller intervals and perform a separate analysis within each interval. Suppose that the cases of interest are those incident during one such short time period, from t_1 to t_2, and that a control group is sampled from those who are at risk of becoming a case during that period. Implicit here is that individuals who became cases in the past are not part of the underlying population at time t_1, for example those with a prevalent condition. Formally, this achieves the aim of comparing the cases with a group of individuals who were 'at risk' of becoming a case, that is, who were eligible to become a case but who did not do so in that interval. The idea of splitting the time into subintervals can be taken to the limit in which that controls are matched to cases in continuous time, by choosing individuals who were at risk of becoming a case at each event time. This approach to control selection naturally allows for a changing underlying population and for exposures that may change over time. The explicit incorporation of *event times* into a case-control study also enables the estimation of *rate ratios*. This method of sampling controls is sometimes called *incidence density sampling* and we will discuss it further in Chapters 2 and 3 and in detail in Chapter 7.

1.4.3 Case-control-within-cohort studies

Case-control sampling designs that take account of time are well suited to use in cohort studies, where the underlying population is an assembled cohort being followed prospectively, often for a range of outcomes, over many years. Although some prospective studies collect full exposure and intrinsic variable information on all participants, it is often necessary to measure certain expensive variables only within subsets of the cohort.

The two main designs for appropriate sampling within assembled cohorts, taking account of time, are the nested case-control design and the case-subcohort design. In both these designs cases are compared with a subsample of individuals from the risk set at their event time. The nested case-control design is the continuous-time version of the incidence density sampling approach described above. The general case-subcohort design extends the special case outlined in Section 1.4.1 (see Table 1.3) to a situation which incorporates event times. Nested case-control studies are discussed in detail in Chapter 7, and case-subcohort studies in Chapter 8.

1.5 Further details on the sampling of cases and controls

1.5.1 Identifying the cases

The first step in conducting a case-control study is to define carefully the specific outcome that defines a case and then to obtain an appropriate set of cases.

One possibility is that the investigator has in mind a particular underlying population of interest, for example the set of individuals living in a particular geographical area during a specified, usually relatively short, time interval. In most settings it is likely that the underlying population is intended to represent a wider population still, for example some subset of the UK population. Then it would be hoped that all, or at least a random sample of, the cases occurring within the underlying population could be identified, and the relevant exposure and other information obtained from them. This type of case-control study is referred to as *population based* and sometimes the well-defined population in which the cases occur is called a *primary base*. A special situation arises when the case-control study is nested within a prospective cohort, as noted in Section 1.4.3.

In another situation, cases are selected from a particular, convenient, source. For example, where the outcome is disease diagnosis the cases may be individuals attending one hospital over a specified time period. The population in which the cases arose is then somewhat hypothetical in that the complete set of individuals forming that population may not be well defined or enumerated. When the underlying population is of this type it is sometimes referred to as a *secondary study base*. In the above example the underlying population might be considered to be the population of individuals who were living within the catchment area of the hospital during the time period in question and who would have attended that hospital if they had become cases.

In a population-based case-control study it may be difficult to be certain that all cases that occur within the specified population have been identified, or alternatively that a random sample of such cases has been obtained. In the situation of a case-control study with a secondary base, the identification of cases is by definition straightforward.

Case-control studies typically select as cases individuals or units which have been identified as 'new' cases, in the sense that the outcome of interest has occurred recently for that individual. These are described as *incident* cases. Incident cases arise when a source of cases is monitored prospectively, for example a hospital or workplace. The selection of individuals remains retrospective and on the basis of their outcomes. Less often, and in general less reliably, cases may be identified historically from individuals or units who had the outcome some time ago. These are described as *prevalent* cases. The use of prevalent cases may be risky, especially in a medical context, because individuals who have had a particular outcome, e.g. disease diagnosis, and survived, so that they can be sampled for the study some time later, may differ in some systematic way from cases arising in the population.

1.5.2 Sampling controls

It was noted above that sampling a suitable group of cases may be difficult in a population-based study. However, in this situation the pool of potential controls is likely to be well defined and the selection of controls by random sampling from the study base is, at least in principle, straightforward. Methods for sampling controls in a population-based case-control study include identifying individuals using an electoral roll or information from a census, or, perhaps, if there is no such suitable list, from those who live in the same neighbourhood as the cases, for example in the same street. Another method used historically is random digit dialling, in which individuals are contacted by randomly selecting telephone numbers from telephone exchanges relevant to the study base. In the special situation of a case-control study nested within a prospective cohort study, the underlying population is formally assembled and control selection is straightforward. As discussed in Section 1.4, time may have to be incorporated in some settings.

When cases are selected from a secondary base, the sampling of controls is more difficult and care is needed in defining the underlying population in such a way that controls are relevant and can be suitably selected. It is generally not feasible to assemble a control group by random sampling from a secondary study base, because it is not enumerated; instead controls are sampled from an available source of individuals who are easily contactable and for whom exposure information can be obtained. When cases are obtained from a hospital, controls could be sampled from the set of individuals admitted to the same hospital in which the cases were observed for reasons unrelated to the outcome of interest but during the time period in which cases were identified. Other methods used to select controls from a secondary study base include sampling individuals who attend the same general medical practices as the cases, who live in the same neighbourhoods as the cases, as in the primary base example, or who are friends or relatives nominated by the cases.

So far we have discussed the sampling of controls from the underlying population. Controls can also be selected using some modification of this, such as, for example, stratified random sampling or frequency matching, or using individually matched controls as discussed in Section 1.3.

If the sampling of controls is even indirectly associated with the exposure, that is, the controls differ systematically from the cases in an unknown way that is associated with the exposure, then inferences about the association between exposure and outcome will be systematically distorted. This issue is discussed in Section 1.6.

1.6 Bias in case-control studies

1.6.1 Preliminaries

A number of potential sources of bias are specific to some case-control study designs. These are of broadly two types, one concerned with the choice of individuals for study and one with the measurement of exposure on individuals already selected. We discuss these in turn.

1.6.2 Selection bias

Bias can arise if controls are sampled in such a way that they do not represent the intended underlying population of interest in some unknown respect that therefore cannot be corrected in analysis. This is referred to as *selection bias*. One way in which selection bias can occur is if the distribution of exposure in the sampled controls differs systematically, in an unknown way, from that in the pool of potential controls in the population from which the cases arose. This is illustrated in Figures 1.6(a), (b), where X has a direct arrow to the indicator of selection D or an arrow to D via another variable. Selection bias can be viewed as a situation in which both the exposure and the outcome have a common effect, D. Conditioning on a common effect of exposure and outcome induces a relationship between X and Y even if X does not affect Y. If there is an effect of X on Y then conditioning on the common effect D will result in a biased estimate of the effect.

Figure 1.6(c) illustrates another way in which selection bias can arise. Here, unmeasured variables W_U affect both exposure X and selection D. The distribution of unobserved intrinsic variables therefore differs systematically in the selected controls compared with the potential controls. If the W_U had been observed then conditioning on W_U would have been a way of correcting for the selection bias. Here selection depends on intrinsic variables, adjustment for which would allow estimation of the effects of primary interest.

Example 1.1: Selection bias To outline a situation in which selection bias can occur, consider the following artificial example. The cases are individuals being treated following a particular cancer diagnosis, Y, in a hospital during a short time period; the exposure of interest, X, is smoking status. Suppose that in reality smoking has no effect on this type of cancer. Now suppose that controls are sampled from individuals who have been admitted to the hospital during the relevant time period and that, for convenience,

(a)

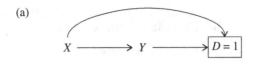

(b)

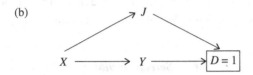

(c)

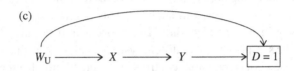

Figure 1.6 Selection bias in case-control studies. Exposure is
denoted by X and outcome by Y. The binary variable D is an
indicator, taking the value 1 for individuals in the underlying
population who are selected for the case-control study. (a)
Selection depends in some way directly on exposure; (b) selection
depends on exposure through another variable J; (c) both
exposure and selection depend on unobserved variables W_U.

they are selected from a ward on which individuals are treated for respiratory illness. We assume that smoking is associated with an increased risk
of respiratory illness. The selection of controls from those with respiratory
illness has the result that smokers are over-represented in the control sample compared with the smoking distribution in the study population with
$Y = 0$. Then, given their selection to the case-control study, controls are
more likely than cases to be smokers. Hence conditioning on $D = 1$ results
in an apparent reduced risk of the outcome $Y = 1$ among smokers. This
situation is as in Figure 1.6(a) but with the arrow from X to Y removed.

Thus, the use of hospital controls relies on the assumption that the exposure is independent of the reasons for admission to hospital in the control
group.

1.6.3 Bias due to retrospective exposure ascertainment

The retrospective collection of exposure information is another, quite different, source of potential bias. Bias arises if some aspect of the data

collection differs systematically between cases and controls, that is, if the case or control status affects some aspect of the exposure measurement. Some examples of how this can occur are as follows.

- *Recall bias* refers to the differential recall of exposure information by cases and controls. It can arise if cases are more likely than controls to remember certain past events or exposures. The relative accuracy of exposure information could also differ between cases and controls. Differential recall or accuracy of recall can also occur in situations where proxies, for example next of kin, are used to obtain exposure information for cases who may be dead or too ill to respond. This is a particular problem if prevalent rather than incident cases are used. In a medical context, the retrospective collection of exposure information would be a cause for concern if the outcome affects memory or cognition.
- Bias can originate also from an interviewer who is obtaining exposure information if he or she is aware of the case or control status of the interviewee and this affects his or her recording of the exposure information; this is sometimes referred to as *information bias*. It can be avoided by blinding interviewers to the participants' diagnoses when possible.
- In some settings the collection of exposure information may involve biological measurements obtained retrospectively and it is possible that the outcome could affect the exposure measurement either via biological mechanisms, in the case of disease outcomes, or by changes in habits as a result of the outcome or changes leading up to the outcome. In this situation, an association between exposure and outcome might involve a reversal of the usual causal process. Again, this may be a particular cause for concern in studies using prevalent cases.

When a case-control study is nested within a larger investigation, exposure information may have been obtained prospectively but is only fully processed, for example questionnaire responses computerized or stored biological samples analysed, within the case-control sample. In this situation the issues discussed above are largely avoided.

Many studies, whether prospective or retrospective, suffer potential bias due to errors of measurement. In a range of settings, the exposures of interest are subject to measurement error. The retrospective collection of exposure information in some case-control studies leads to greater potential for differential errors in the measurements, that is, errors that differ according to case or control status. This possible bias may be more important than random errors. We return to the issue of measurement errors in Chapter 9.

1.7 Use of case-control studies

Case-control studies have a long history in epidemiology and in medical studies more generally, with applications ranging from the study of outbreaks of infectious disease to the investigation of risk factors for cancer or heart disease. In this context exposures of interest may be lifestyle factors such as tobacco or alcohol use, occupational exposures such as shift work or contact with a chemical or biological measures such as blood pressure or cholesterol level.

Case-control type studies are used also in econometrics, where they are called *choice-based sampling*, the term arising from an application in transport economics in which factors associated with alternative modes of transport to work, some rarely used, were assessed. Other examples of outcomes of interest in econometrics are loan defaults and household bankruptcy, where risk factors may relate to household or personal debt, assets and sociological factors. In political science case-control studies have been used to investigate factors associated with wars between pairs of countries and, in sociology, to study the association between drug use and homicide and the association of career choice in young people with educational and family background and psychological traits.

Below we give some outline examples.

Example 1.2: Case-control study with secondary base   An early and influential use of a case-control study was that in which Doll and Bradford Hill (1950) investigated the association between smoking and lung cancer. The objective of the study was 'to determine whether patients with carcinoma of the lung differed materially from other persons in respect of their smoking habits or in some other way which might be related to the atmospheric pollution theory'. A set of cases was obtained by asking 20 London hospitals to report all patients with carcinoma of the lung between April 1948 and October 1949. Patients were interviewed at the hospital by one of four almoners and, for each case, the almoner interviewed a patient of the same sex, within the same five-year age group and in the same hospital at around the same time. The study excluded individuals aged over 75 years and some patients were also excluded because they could not be interviewed for one reason or another, including that they had been discharged, were too ill or were dead. After exclusions, the study comprised 709 lung cancer cases and 709 matched controls. Study participants were questioned in detail about the history of their smoking habits. The case-control design allowed the authors to compare smoking habits in case and control groups which were comparable in terms of sex, age and region of residence.

Doll and Bradford Hill (1950) asked whether their results could 'be due to an unrepresentative sample of patients with carcinoma of the lung or a choice of control series which was not truly comparable' and 'could they have been produced by an exaggeration of their smoking habits by patients who thought they had an illness which could be attributed to smoking? Could they be produced by a bias on the part of the interviewers in taking and interpreting the histories?'

The authors performed a number of investigations into potential sources of bias, and we mention just two of these here. Representativeness of cases and controls was considered: no reason was found to believe the cases might be unrepresentative of lung cancer patients attending London hospitals, and there was no evidence to suggest that the hospital controls smoked less than average, which would have biased the results; it was shown that the level of smoking was similar for all disease groups used in the control group. Interviewers could not be made blind to case or control status, so information bias was possible. This was assessed by looking at the recorded smoking habits of individuals interviewed as cases but who turned out not to have lung cancer. Their recorded smoking habits did not differ significantly from those of the controls.

Example 1.3: Population-based case-control study   Chen *et al.* (1998) used a population-based case-control study to investigate the association between the use of sunlamps and melanoma. The underlying population was that of Caucasians resident in Connecticut, USA, between 15 January 1987 and 15 May 1989. All primary cases of melanoma in this population were identified using the Rapid Case Ascertainment System at the Cancer Prevention Research Unit at Yale University. This yielded 650 cases, 624 of whom provided complete information on the exposures relating to sunlamp use and were included in the final study. Controls were sampled from the Connecticut general population using random digit dialling and were frequency matched to the cases by sex and age distribution. This gave 512 controls after exclusion of those who could not be interviewed, those with incomplete exposure information and non-Caucasians.

Example 1.4: Case-control study in social science   Recchi (1999) used a case-control study in an investigation of determinants of career choice in young people in Italy, in particular focusing on the choice to become a political party activist ('cases'), this commonly being the first step to a political career. The cases were 115 political party activists identified from party youth organizations in 1992. The controls were a random sample of

322 non-activists from the same age cohort as the cases. The case and control groups were interviewed to obtain information on a range of individual characteristics and experience. The exposures considered were 'sense of group belonging', 'sociability' and 'personal control', and the intrinsic variables accounted for included sex, parental class, urban residence and work-force participation.

Example 1.5: Case-control study in econometrics   A case-control study, or choice-based sample, was used by Domowitz and Sartain (1999) to investigate risk factors for the decision to file for bankruptcy in the USA. The outcome in this study was household bankruptcy, and the study distinguished between the two types of bankruptcy that could be chosen under the US Bankruptcy Code. A set of 827 cases was randomly sampled from 17 565 cases of bankruptcy in five court districts in the second and third quarters of 1980. The exposure variables of interest included whether medical debt exceeded 2% of income, homeownership, marital status, ratio of total debt and total household assets and credit card debt. For cases this information was obtained from the bankruptcy petitions. A control set of 1862 households was obtained from the 1983 Survey of Consumer Finances, the only source of detailed information on assets and debts. The authors noted that there was no rewriting of bankruptcy law or significant trends in household behaviour relating to assets and debts during the period 1980–3.

Example 1.6: Case-control study in political science   King and Zeng (2001b) described the use of a case-control design to study factors associated with state failures, defined as a 'complete or partial collapse of state authority'. The study was part of the work of the State Failure Task Force in the United States. From 1955 to 1998, 127 state failures ('cases') were identified in within a basis of 195 different countries. For each state failure, three countries that did not have a state failure in that year were sampled. A large amount of time-dependent information was available for each country, for example on population size, democracy, trade openness and infant mortality. The investigators studied associations between many factors and the outcome, with risk factors of interest being those measured two years prior to the time of the state failure. Although King and Zeng (2001b) identified some possible errors in the treatment of the data, this study illustrates the potential for using case-control studies in a variety of fields.

Example 1.7: Sampling of controls over time   In a veterinary study of bovine TB by Johnston *et al.* (2011) the husbandry practices of farms experiencing a case of TB in a particular calendar year were compared with those of control farms in the same area that were free of TB in the study period up to the time of the case. Some of these control farms subsequently experienced TB during the study period. There would, however, have been a serious risk of bias if controls had been restricted to farms with no TB over the whole study period. These would have tended to be farms with a small number of animals and possibly atypical also in various unknown but relevant respects.

Example 1.8: Bias due to retrospective exposure ascertainment   Samanic *et al.* (2008) used a case-control study to investigate the association between exposure to pesticides and brain cancer. Between 1994 and 1998, a set of 686 newly diagnosed cases of adult brain cancers (glioma and meningioma) was identified in three US hospitals specializing in brain tumour treatment. A set of 799 controls was sampled from patients admitted to the same hospitals for unrelated conditions, using frequency matching by sex, 10-year age group, race or ethnicity, hospital and distance of residence from hospital. Participants or proxies, their next of kin, were interviewed to obtain occupational histories and information relating to occupational exposure to pesticides.

In this study there was a possibility of differential recall of occupational exposure information by glioma cases and meningioma cases, because glioma patients can experience impaired memory and cognition to a greater degree than meningioma cases. Bias could also arise because for some study participants exposure information was obtained from a proxy. The possibility of differential recall by participants and proxies was investigated by repeating the analyses while excluding individuals for whom exposure information was obtained by proxy, but there was no significant change in the results. A further potential source of information bias was ruled out by blinding interviewers to the participants' diagnoses.

Notes

Section 1.1. There is an extensive literature on the design and analysis of retrospective case-control studies. A general overview was given by Wacholder and Hartge (1998). The main text is that by Breslow and Day (1980), which focuses on the use of case-control studies in cancer epidemiology. Cole's introduction to that book provides an overview of the use of case-control studies. Breslow (1996) gives an overview of case-control designs and methods of analysis.

Section 1.2. Cornfield (1951) showed that the odds ratio from a 2 × 2 table such as Table 1.1 is the same whether the data was collected retrospectively or prospectively; he used this in case-control studies of cancers of the lung, breast and cervix. Cornfield (1951) also pointed out that the odds ratio provides an estimate of the risk ratio when the case outcome is rare. Edwards (1963) showed a uniqueness aspect of the odds ratio, that is, the odds ratio is the same retrospectively as it is prospectively and any other measures with this property are functions of the odds ratio. Justifications of the use of logistic regression to estimate the odds ratio were given by Prentice and Pyke (1979) and also by Farewell (1979) and Anderson (1972).

Section 1.3. An early discussion of the use of matching in retrospective studies is that of Mantel and Haenszel (1959).

Section 1.4. Different approaches to control sampling in case-control studies, and discussions of which quantities can be estimated, were described by Rodrigues and Kirkwood (1990), Pearce (1993), Greenland and Thomas (1982), Miettinen (1976, 1985) and Langholz (2010). See Rothman *et al.* (2008, Chapter 8) for a summary.

Section 1.5. The definition of the underlying population and the choice of cases and the sampling of controls in epidemiological case-control studies was discussed in detail by Wacholder *et al.* (1992a, b, c) and also by Gail (1998a, b) and Wacholder and Hartge (1998).

Section 1.6. See Hernán *et al.* (2004) and Didelez *et al.* (2010) for discussions of selection bias in case-control studies and its graphical representation. Hernán *et al.* (2004) use in causal diagrams a box around variables on which there is to be conditioning; see Figures 1.3 and 1.4.

Section 1.7. The vast majority of the literature on case-control studies has a medical or epidemiological context. However, King and Zeng (2001a, c) give a guide to the use of case-control studies in a political science setting, and the book by Manski and McFadden (1981) discusses the use of choice-based samples in econometrics. See Xie and Manski (1989) for the use of case-control studies in sociology, where such studies are sometimes described as response-based sampling. Lilienfeld and Lilienfeld (1979) and Paneth *et al.* (2002a, b) have given histories of the use of case-control studies.

2

The simplest situation

- The cornerstone of the analysis of case-control studies is that the ratio of the odds of a binary outcome Y given exposure $X = 1$ to that given $X = 0$ is the same as the ratio of the odds where the roles of Y and X are reversed. This result means that prospective odds ratios can be estimated from retrospective case-control data.
- For binary exposure X and outcome Y there are both exact and large-sample methods for estimating odds ratios from case-control studies.
- Methods for the estimation of odds ratios for binary exposures extend to categorical exposures and allow the combination of estimates across strata. The latter enables control for confounding and background variables.
- For binary exposure X and outcome Y, the probabilities of X given Y and of Y given X can be formulated using two different logistic regression models. However, the two models give rise to the same estimates of odds ratios under maximum likelihood estimation.
- Rate ratios can be estimated from a case-control study if 'time' is incorporated correctly into the sampling of individuals; a simple possibility is to perform case-control sampling within short time bands and then to combine the results.

2.1 Preliminaries

Many central issues involved in the analysis of case-control data are illustrated by the simplest special case, namely that of a binary explanatory variable or risk factor and a binary outcome or response. For this discussion we use a notation consistent with that widely used for statistical studies of dependence, namely taking X for a random variable representing the *exposure*, or explanatory feature, and Y for a random variable representing the *outcome* or response. Here both are binary, taking values 0 and 1. For a tangible interpretation for X we may think of 0 as representing no exposure to a risk factor and 1 as exposure. For Y we may think of 0 as representing no disease or adverse outcome and 1 as representing an adverse outcome.

Under this interpretation, those with outcome $Y = 1$ are referred to as *cases*. This notation extends fairly easily to more general situations.

We consider initially a population in which each study individual falls into one of four types indexed first by exposure status and then by outcome status. The study population thus defines probabilities

$$\pi_{00}, \quad \pi_{01}, \quad \pi_{10}, \quad \pi_{11}, \tag{2.1}$$

where

$$\pi_{xy} = \Pr(X = x, Y = y), \tag{2.2}$$

and

$$\pi_{00} + \pi_{01} + \pi_{10} + \pi_{11} = 1. \tag{2.3}$$

These probabilities are shown in Table 2.1(a).

From the probabilities π_{xy} we may calculate, in particular, two sets of conditional probabilities: the probability of outcome Y conditional on exposure X and the probability of exposure X conditional on outcome Y. The two possibilities correspond to two ways in which the population in question may be investigated. These are, first, from a prospective standpoint in which sampling from the population is based on initially recording the exposure or, second, from a retrospective or case-control standpoint, in which the sampling is based on the outcome.

In a prospective or cohort study we take a random sample from the population. More generally, we might take random samples independently from the two subpopulations with exposures $X = 0$ and $X = 1$. In both situations we then observe the corresponding values of Y. From this information we may estimate and compare the probabilities

$$\Pr(Y = y|X = x) = \pi_{xy}/(\pi_{x0} + \pi_{x1}), \tag{2.4}$$

for $x, y = 0, 1$. These are shown explicitly in Table 2.1(b).

As discussed in Chapter 1, the motivation for a case-control study comes primarily from situations in which outcome status $Y = 1$, say, is a rare event, so that very large sample sizes become necessary if prospective studies are to include even a modest number of such individuals. This situation motivates sampling independently the two subpopulations defined by outcome status $Y = 0$ (the controls) and outcome status $Y = 1$ (the cases). The exposure status is then observed for the sampled individuals. Precisely how the sampling is done is critical; some issues relating to the retrospective sampling of cases and controls were discussed in Chapter 1. In the present chapter we focus primarily on the form of case-control study

Table 2.1 *Probabilities associated with binary explanatory and binary response variables sampled in various ways.*

(a) The underlying population structure.

	$Y = 0$	$Y = 1$
$X = 0$	π_{00}	π_{01}
$X = 1$	π_{10}	π_{11}

(b) Prospective sampling: separate samples from subpopulations $X = 0, 1$ with relevant conditional probabilities.

	$Y = 0$	$Y = 1$	$\Pr(Y = y \mid X = x)$
$X = 0$	π_{00}	π_{01}	$\pi_{01}/(\pi_{01} + \pi_{00})$
$X = 1$	π_{10}	π_{11}	$\pi_{11}/(\pi_{10} + \pi_{11})$

(c) Retrospective (case-control) sampling: separate samples from subpopulations $Y = 0, 1$ with relevant conditional probabilities.

	$Y = 0$	$Y = 1$
$X = 0$	π_{00}	π_{01}
$X = 1$	π_{10}	π_{11}
$\Pr(X = x \mid Y = y)$	$\pi_{10}/(\pi_{10} + \pi_{00}) = \theta_0$	$\pi_{11}/(\pi_{11} + \pi_{01}) = \theta_1$

in which controls are selected from individuals who do not have the event during the period over which cases are observed. Extensions to incorporate time into the selection of controls are considered in Section 2.10 and later, in more detail, in Chapters 7 and 8.

The above discussion leads us to consider the conditional probabilities for exposure X given outcome Y, namely

$$\Pr(X = x \mid Y = y) = \pi_{xy}/(\pi_{0y} + \pi_{1y}).\tag{2.5}$$

These probabilities are shown in Table 2.1(c) and play a central role in the discussion. It is convenient to write, for the probability of exposure $X = 1$ given outcome status $Y = y$,

$$\theta_y = \Pr(X = 1 \mid Y = y).\tag{2.6}$$

2.2 Measuring association

Associations between two binary variables, arranged as in Table 2.1, can be assessed by a comparison of the observed frequencies of individuals in each of the four table cells with the expected frequencies under the hypothesis that X and Y are independent, that is, that the proportion of people with $Y = 1$, say, does not differ according to X and vice versa. The comparison can be formalized using a chi-squared test. However, this treats the two variables on an equal footing and, for studies of relationships between exposures and outcomes, it is of interest to go further.

In a study of the association between the exposure X and the outcome Y, as defined above, we are usually interested in how the prevalence of the outcomes $Y = 0$ and $Y = 1$ differs among individuals with different exposures X. For example, in a study of the association between smoking and lung cancer we are interested in whether the proportion of lung cancer cases ($Y = 1$) that arise in a group of smokers ($X = 1$) differs from the proportion of lung cancer cases arising in a group of non-smokers ($X = 0$). Therefore, for interpretation we would like to compare the prospective conditional probabilities $\Pr(Y = 1 | X = 1)$ and $\Pr(Y = 1 | X = 0)$. The comparison could be via the ratio of the two probabilities, that is,

$$\frac{\Pr(Y = 1 | X = 1)}{\Pr(Y = 1 | X = 0)}. \tag{2.7}$$

This ratio is called a *risk ratio*, or a *relative risk*. Using the notation shown in Table 2.1(b) the risk ratio is

$$\frac{\pi_{11}/(\pi_{10} + \pi_{11})}{\pi_{01}/(\pi_{00} + \pi_{01})}. \tag{2.8}$$

However, from Table 2.1(c) it is evident that the risk ratio cannot be calculated from a case-control study of the form considered here without additional information.

Instead of the risk ratio we consider comparing the two probabilities $\Pr(Y = 1 | X = 1)$ and $\Pr(Y = 1 | X = 0)$ using an *odds ratio*. To do this we first define the *odds* of an arbitrary event A with probability $\Pr(A)$ to be the ratio $\Pr(A)/\{1 - \Pr(A)\}$. The odds of the outcome $Y = y$ given exposure $X = x$ are therefore

$$\frac{\Pr(Y = y | X = x)}{1 - \Pr(Y = y | X = x)}, \tag{2.9}$$

which is often best recorded on a log scale.

Note that for a binary random variable, such as Y, taking values 0 and 1, a consequence of interchanging 0 and 1 is a change in the sign of the log odds. This has the formal consequence that many analyses suggested here and later are unchanged, except for sign, by interchanging the labelling of the values of the binary variables concerned.

It is again evident that these odds can be calculated from a prospective study but not from a retrospective, or case-control, study. The cornerstone of the analysis of case-control studies is that the *ratio* of the odds of $Y = 1$ given $X = 1$ and of $Y = 1$ given $X = 0$ calculated from the prospective study is the same as the formally corresponding ratio of odds in the case-control study of $X = 1$ given $Y = 1$ to that given $Y = 0$. In fact

$$\frac{\Pr(Y = 1|X = 1)/\Pr(Y = 0|X = 1)}{\Pr(Y = 1|X = 0)/\Pr(Y = 0|X = 0)}$$
$$= \frac{\Pr(X = 1|Y = 1)/\Pr(X = 0|Y = 1)}{\Pr(X = 1|Y = 0)/\Pr(X = 0|Y = 0)}. \tag{2.10}$$

That is, the ratio of the odds of the outcome $Y = 1$ given exposure $X = 1$ and the corresponding odds given $X = 0$ can be found from the case-control data by in effect *pretending* that $X = 1$ is a response to explanatory feature $Y = 1$, that is, by interchanging the roles of the explanatory and response variables. Using the notation in Table 2.1, the odds ratio can be written as

$$e^{\psi} = \frac{\pi_{11}\pi_{00}}{\pi_{10}\pi_{01}}, \tag{2.11}$$

where ψ is the log odds ratio. Note that this is the same as the cross-product ratio derived from the 2×2 table of probabilities in Table 2.1(a). The ratio (2.11) is widely used as an index of association in 2×2 tables in which both variables are treated on an equal footing.

Now using the notation introduced for the case-control study in Table 2.1(c), the odds ratio can also be written as

$$e^{\psi} = \frac{\theta_1/(1 - \theta_1)}{\theta_0/(1 - \theta_0)}. \tag{2.12}$$

In many applications of case-control studies the probabilities of $Y = 1$, say, are small; indeed, as we have seen, this is the primary motivation of such a study design. Then the probability $\Pr(Y = 0|X = x)$ is to a close approximation equal to unity. Thus the odds ratio in (2.10) is to a close approximation equal to the ratio of probabilities $\Pr(Y = 1|X = 1)/\Pr(Y = 1|X = 0)$, that is, the risk ratio. Except possibly for those with a strong gambling background, the risk ratio may be more easily interpreted than the odds ratio.

Table 2.2 *Summary of data from a case-control study of n individuals; $r. = r_0 + r_1$.*

	$Y = 0$	$Y = 1$	Total
$X = 0$	$n_0 - r_0$	$n_1 - r_1$	$n - r.$
$X = 1$	r_0	r_1	$r.$
Total	n_0	n_1	n

In Chapter 8 it will be shown that risk ratios can be estimated directly from a special type of case-control study without any need for an assumption that cases are rare in the underlying population. In Section 2.10 it will be shown that a different measure of association, the *rate ratio*, can be estimated from a case-control study when time is incorporated into the selection of controls. In general, what can be estimated from a case-control study depends on how the controls are selected and then on correct incorporation of the control sampling into the analysis. When cases are rare, the different ratios will be nearly equal.

2.3 Statistical analysis

2.3.1 Estimating the odds ratio

In this section we outline methods for estimating the odds ratio using data from a case-control study. Suppose now that observations are available on n mutually independent individuals. For the analysis of a case-control study we suppose initially that n_1 individuals, the cases, are selected at random from the subpopulation with $Y = 1$ and n_0 individuals, the controls, are selected at random from the subpopulation with $Y = 0$, generating the data shown in Table 2.2. The independence of individuals precludes the matching of individual cases with controls; this requires an extended analysis given in Chapter 3. Often by design n_0/n_1 is a small integer, in particular 1, the latter corresponding to the choice of one control per case. In many applications all available eligible cases in the source population will be used.

We first describe an 'exact' method for the estimation of odds ratios. The frequencies r_1, r_0 in Table 2.2, which are the numbers of exposed cases and controls respectively, are values of independent binomially distributed random variables. This leads to the joint probability

$$\Pr(R_1 = r_1, R_0 = r_0) = \binom{n_1}{r_1} \theta_1^{r_1}(1 - \theta_1)^{n_1 - r_1} \binom{n_0}{r_0} \theta_0^{r_0}(1 - \theta_0)^{n_0 - r_0},$$

$$(2.13)$$

where θ_1 is the probability of exposure $X = 1$ for a case and θ_0 is that for a control, as defined in (2.6).

It is the comparison of the proportions of exposed individuals in the case group and in the control group that informs us of a possible association between exposure and outcome. It is convenient to begin the formal analysis by considering the null hypothesis, that of no difference between the case and control groups in terms of exposure: $\theta_1 = \theta_0$. A sufficient statistic for the unknown common value is the total frequency $R_. = R_0 + R_1$, which is a binomially distributed random variable of index $n = n_1 + n_0$. We therefore condition on its observed value as the only way to obtain a test distribution not depending on a nuisance parameter. The test statistic may be either r_1 or r_0; we will choose the former. Then, under the null hypothesis, the corresponding random variable R_1 has the hypergeometric distribution derived by dividing (2.13) by the marginal binomial probability for $R_.$:

$$\Pr(R_1 = r_1 | R_. = r_.) = \frac{\binom{n_1}{r_1}\binom{n_0}{r_0}}{\binom{n}{r_.}}. \tag{2.14}$$

The same argument shows that for a non-null situation the conditional probability has the generalized hypergeometric form

$$\Pr(R_1 = r_1 | R_. = r_.) = \frac{\binom{n_1}{r_1}\binom{n_0}{r_0} e^{r_1 \psi}}{\sum_k \binom{n_1}{k}\binom{n_0}{r_. - k} e^{k\psi}}, \tag{2.15}$$

where

$$e^\psi = \frac{\theta_1/(1 - \theta_1)}{\theta_0/(1 - \theta_0)} \tag{2.16}$$

is the odds ratio.

The maximum likelihood estimate for the log odds ratio is obtained by equating to zero the derivative with respect to ψ of the log of the likelihood (2.15). The estimated odds ratio $\hat{\psi}$ is therefore obtained by solving

$$r_1 - \frac{\sum_k k \binom{n_1}{k}\binom{n_0}{r_. - k} e^{k\hat{\psi}}}{\sum_k \binom{n_1}{k}\binom{n_0}{r_. - k} e^{k\hat{\psi}}} = 0. \tag{2.17}$$

Special numerical methods are required to obtain the maximum likelihood estimate for the log odds ratio and to obtain confidence intervals or perform tests of significance. In practice, except possibly when one of the frequencies is very small, it is more insightful to use approximate large-sample results based on asymptotic statistical theory and in any case this is often almost unavoidable in more complicated situations. Here, and throughout, by large-sample methods we mean procedures whose theory is based on the limiting behaviour as the amount of information increases indefinitely and which are then used to provide confidence intervals and other procedures with approximately the desired properties. The limiting operation considered in the mathematical theory of such procedures is just a technical device for obtaining approximations, whose adequacy has always in principle to be considered. In the applications considered in this book the approximate procedures are likely to be adequate provided that there are no very small frequencies having a substantial influence on the apparent conclusions. See the appendix for more details.

We now outline a simpler approach to the more formal statistical inference. Conditionally on n_1 and n_0 the frequencies in Table 2.2 follow binomial distributions and the ratios r_1/n_1 and r_0/n_0 provide unbiased estimates of the conditional probabilities θ_1 and θ_0 respectively. It follows that a simple, and intuitive, estimate of the log odds ratio defined below in (3.9) is

$$\hat{\psi} = \log \frac{r_1/(n_1 - r_1)}{r_0/(n_0 - r_0)}, \tag{2.18}$$

always supposing that none of the four frequencies is very small.

By considering a Taylor expansion for $\log(r_1/n_1)$ about θ_1, it can be shown that, given n_1, the asymptotic mean of $\log(r_1/n_1)$ for large n_1 is $\log \theta_1$. A similar result holds for $\log(r_0/n_0)$, so that (2.18) is an asymptotically unbiased estimate of the log odds ratio. This estimate may be called the *empirical estimate*.

2.3.2 Estimating uncertainty

Under the exact method for odds ratio estimation described in the above section, the upper and lower limits in a $(1 - \alpha)$-level two-sided exact equitailed confidence interval for the log odds ratio estimate are given by the values of ψ which satisfy respectively

$$\Pr(R_1 \leq r_1 | R_{.} = r_{.}) = \alpha/2, \quad \Pr(R_1 \geq r_1 | R_{.} = r_{.}) = \alpha/2.$$

The asymptotic variance of the empirical log odds ratio estimate in (2.18) can be shown to take a simple form. It is obtained from the properties of the number of exposed cases, taken conditionally on the total number of cases, noting that the random variable R_1 is binomially distributed since it is the number of successes in n_1 Bernoulli trials. Similarly, the number of exposed controls, R_0, conditional on the total number of controls, is, independently of R_1, binomially distributed since it is the number of successes in n_0 Bernoulli trials.

We may now apply a standard method of local linearization, often called the *delta method*, to derive the asymptotic standard error of the log odds and hence of the log odds ratio.

Partly to illustrate a more widely applicable approach and partly to explain the structure of the resulting formula, we shall use a less direct approach. The general idea of this is to relate the comparison of frequencies associated with binary variables to comparisons of the logs of *independent* Poisson-distributed random variables, for which a simple calculation of asymptotic variances is possible. The argument is fairly general and has implications for the use of log Poisson regression for the analysis of case-control and other data. The essence is contained in the following series of results.

Result 1. If two random variables V_0, V_1 have independent Poisson distributions with means γ_0, γ_1 then, conditionally on $V_1 + V_0 = v$, the random variable V_1, say, has a binomial distribution of index v and probability

$$\frac{\gamma_1}{\gamma_1 + \gamma_0}. \tag{2.19}$$

The proof is by direct calculation using the property that $V_1 + V_0$ has a Poisson distribution of mean $\gamma_1 + \gamma_0$. A more general result is that if $V_0, \ldots, V_p$ have independent Poisson distributions with means $\gamma_0, \ldots, \gamma_p$ then, conditionally on $V_0 + \cdots + V_p = v$, the V_j have a multinomial distribution corresponding to v trials with cell probabilities $\gamma_j / \sum_k \gamma_k$.

Result 2. Next, if a random variable V has a Poisson distribution with mean γ then, for large γ, $\log V$ is asymptotically normal with mean $\log \gamma$ and variance $1/\gamma$, estimated by $1/V$. In a slight abuse of notation we say that $\log V$ has asymptotic standard error $1/\sqrt{V}$, meaning that as γ tends to infinity the distribution of the random variable

$$\frac{\log V - \log \gamma}{\sqrt{(1/V)}} \tag{2.20}$$

converges to the standard normal distribution. The proof is by local linearization of the function $\log(V/\gamma)$. A more general result is that if $V_0, \ldots, V_p$ have independent Poisson distributions with means $\gamma_0, \ldots, \gamma_p$

then two linear functions

$$\sum_k c_k \log V_k, \quad \sum_k d_k \log V_k$$

have an asymptotic bivariate normal distribution with means

$$\sum_k c_k \log \gamma_k, \quad \sum_k d_k \log \gamma_k$$

and asymptotic variances and covariance

$$\sum_k c_k^2 / V_k, \quad \sum_k d_k^2 / V_k, \quad \sum_k c_k d_k / V_k. \tag{2.21}$$

Result 3. Similarly, the asymptotic covariance between a linear combination $\sum_k c_k \log V_k$ and a linear combination of the V_k themselves, say $\sum_k d_k V_k$, is $\sum_k c_k d_k$. Therefore if $\sum_k c_k \log V_k$ is a *contrast*, that is, if $\sum_k c_k = 0$ then asymptotically

$$\mathrm{cov}\left(\sum_k c_k \log V_k, \sum_k V_k\right) = 0. \tag{2.22}$$

Because the two random variables in (2.22) can be shown to be asymptotically bivariate normally distributed, it follows that they are asymptotically independent. This implies that by extension, when several contrasts of the log V_k are considered, the asymptotic variance matrix of any set of contrasts of the log V_k is the same whether or not it is conditioned on the observed value of the total ΣV_k. It follows that the asymptotic covariance matrix of a set of such contrasts, $\sum_k c_k \log V_k$, is the same in a formulation in terms of binomial or multinomial distributions as it is in the rather simpler Poisson model. Therefore we will use the latter for many of the following calculations.

We now return to estimating the variance of the empirical log odds ratio estimate in (2.18). An application of the above results is used to find the variance of $\log\{R_1/(N_1 - R_1)\}$ given $N_1 = n_1$ and of $\log\{R_0/(N_0 - R_0)\}$ given $N_0 = n_0$. Suppose that R_1 and $N_1 - R_1$ are treated as independent Poisson random variables; then using result 1 above it follows that, conditionally on $N_1 = n_1$, R_1 has a binomial distribution with parameters (n_1, θ_1), which is our originating model. Note next from the second result that $\log R_1$ and $\log(N_1 - R_1)$ are both asymptotically normally distributed. From result 2, it follows that $\log\{R_1/(N_1 - R_1)\}$ is asymptotically normally distributed with mean $\log\{\theta_1/(1 - \theta_1)\}$ and variance $1/(n\theta_1) + 1/\{n(1 - \theta_1)\}$,

Table 2.3 *Summary of the data from a case-control study of sunlamp use and melanoma.*

Ever used a sunlamp?	Controls	Cases	Total
No	417	483	900
Yes	95	141	236
Total	512	624	

which is estimated by $1/r_1 + 1/(n_1 - r_1)$. Result 3 gives

$$\text{cov}\{\log(R_1/(N_1 - R_1)), N_1\} = 0.$$

Hence the asymptotic variance of $\log\{R_1/(N_1 - R_1)\}$ is the same conditionally or unconditionally on the value of N_1. That is, when R is distributed in a binomial distribution with parameters (n, θ), we can calculate the asymptotic variance of $\log\{R/(N - R)\}$, given $N = n$, by treating R and $N - R$ as having independent Poisson distributions and calculating the asymptotic variance of $\log\{R/(N - R)\}$.

From these results it follows directly that the asymptotic variance of the estimated log odds ratio, $\hat{\psi}$, given in (2.18), is

$$1/r_1 + 1/(n_0 - r_0) + 1/(n_1 - r_1) + 1/r_0. \tag{2.23}$$

The local-linearization approach yields the same result.

Example 2.1: Empirical estimation of an odds ratio To illustrate the methods described in this chapter we return to the case-control study by Chen *et al.* (1998) of the association between sunlamp use and the risk of melanoma, which was introduced in Example 1.3; see Section 1.7. The data on sunlamp use among cases and controls is shown in Table 2.3.

The empirical estimate of the log odds ratio is

$$\hat{\psi} = \log \frac{141 \times 417}{95 \times 483} = \log 1.281 = 0.248, \tag{2.24}$$

with asymptotic variance

$$\text{var}(\hat{\psi}) = 1/141 + 1/483 + 1/95 + 1/417 = 0.022\,09, \tag{2.25}$$

and standard error 0.149, giving an approximate 0.95 confidence interval $(-0.043, 0.539)$. An approximate asymptotic 0.95-level confidence interval for the odds ratio is therefore $(0.957, 1.715)$.

An exact method based on the generalized hypergeometric distribution gives an odds ratio estimate of 1.28, with 0.95-level confidence interval (0.948, 1.736).

We calculated the above odds ratios using numbers taken from the paper by Chen *et al.*; the paper itself gives much more detailed results, including adjusted odds ratios.

2.4 Model formulation

In this section we set out explicitly the distinction between three different statistical models associated with the system under study. The reason for formulating the situation in the way described in this section is partly to provide a basis for studying more complicated and realistic situations, though for the discussion here we remain restricted to the special case of a single binary explanatory variable.

- First, we have a *population model*. This describes the joint distribution of the exposure and outcome (X, Y) in the target population being investigated. That is, it specifies the probabilities $\pi_{xy} = \Pr(X = x, Y = y)$ for $x, y = 0, 1$.
- Next, there is the *sampling model*. In a case-control study this specifies the distribution of the exposure X in random samples of sizes n_0 and n_1 drawn from the subpopulations with outcomes $Y = 0$ and $Y = 1$. The sampling model is therefore based on the conditional distribution of X given respectively $Y = 0, 1$.
- Finally, it is helpful for subsequent discussion to introduce a third *formal interpretative* or *inverse model*, specifying here the conditional distribution of Y given $X = 0, 1$, that is, of the response or outcome variable given the explanatory variable. This is the usual specification in studies of dependence. Here, however, it is *not* the distribution generating the data.

The population model, the inverse model and the sampling model refer respectively to Tables 2.1(a), (b), (c).

For some purposes it is helpful to write the sampling model in the form

$$\Pr(X = 1 | Y = 1) = L_1(\psi + \chi), \qquad (2.26)$$
$$\Pr(X = 1 | Y = 0) = L_1(\chi).$$

Throughout it is convenient to take

$$L_1(x) = e^x/(1 + e^x), \quad L_0(x) = 1/(1 + e^x), \qquad (2.27)$$

where $L_1(x)$ is the standardized logistic distribution function. As in the previous notation, ψ denotes the log odds ratio.

For the inverse model we write

$$\Pr(Y = 1 | X = 1) = L_1(\psi + \lambda), \tag{2.28}$$
$$\Pr(Y = 1 | X = 0) = L_1(\lambda).$$

To accord with the previous notation we take

$$e^{\chi} = \pi_{10}/\pi_{00}, \quad e^{\lambda} = \pi_{01}/\pi_{00}, \tag{2.29}$$

and these two quantities are quite different.

For the data specified in Table 2.2 the log likelihood specified by the sampling model is

$$r_1\psi + r_{\cdot}\chi - n_1 \log(1 + e^{\psi + \chi}) - n_0 \log(1 + e^{\chi}). \tag{2.30}$$

It can be shown that the maximum likelihood estimate for the log odds ratio ψ is identical to the empirical estimate in (2.18), with identical asymptotic variance estimate; the latter is found from the observed information, that is, the second derivative of the log likelihood function at the maximum. We refer to the appendix for an outline description of maximum likelihood estimation.

Note that the fitting of the models by maximum likelihood is best achieved by working directly in terms of the parameters of the associated binomial distributions and then reparameterizing into a logistic form. The key facts are that the sampling model and the inverse model yield identical

- maximum likelihood estimates $\hat{\psi}$,
- asymptotic standard errors for $\hat{\psi}$,
- exact inferences for ψ based on the generalized hypergeometric distribution, which is given in (2.15) for the sampling model.

That is, we would make the same inferences about the log odds ratio, ψ, whether the data in Table 2.2 had arisen from a case-control study or from a prospective study. The arguments leading to the generalized hypergeometric distribution are slightly different in the two situations. In the sampling model the numbers of observations on cases and on controls are determined by design, whereas the total number of observations at, say, $X = 1$ is determined by the technical need in frequentist inference to obtain a distribution not depending on a nuisance parameter. In the inverse model the two roles are reversed.

2.5 Design implications for the numbers of cases and controls

An important part of the planning of any investigation involves consideration of the scale of effort involved. In principle this typically involves some balance between the cost of an increased scale of effort and losses due to misleading or incomplete conclusions arising from the imprecision implicit in smallish investigations. It is rare that this balance can be usefully formulated quantitatively. The main design aspect under the control of the investigator is likely to be the number of cases and controls to be selected for a case-control study.

We will use the asymptotic variance of the log odds ratio estimate given in (2.23) to study the choice of the numbers of cases and controls. We consider the precision in the log odds ratio estimate that will be achieved locally near the null hypothesis $\psi = 0$ of no differences in exposure between cases and controls. Under this null hypothesis, whatever the value of Y the distribution of X is the marginal distribution in the population, that is,

$$\Pr(X = x | Y = y) = \Pr(X = x) = \pi_{x0} + \pi_{x1} = \pi_{x.}. \qquad (2.31)$$

It follows from (2.23) that with n_1 cases and $n_0 = cn_1$ controls, taking c controls per case and replacing observed values by their expectations, the large-sample variance of $\hat{\psi}$ is

$$\frac{1 + 1/c}{n_1 \pi_{0.} \pi_{1.}}. \qquad (2.32)$$

The first consequence of (2.32) concerns the role of c, the number of controls per case. Sometimes cases are difficult and/or expensive to obtain, while controls are obtained relatively easily and cheaply. This suggests the use of multiple controls corresponding to each case, the choice of c being in principle an economic one. There are, however, diminishing returns in obtaining additional controls beyond, say, $c = 5$.

The limitation on the size of case-control studies is often primarily in terms of the number of cases involved. A second use of (2.32) is thus in connection with the choice of n_1, the number of cases. The following issues may arise.

- What should be the target number of cases and is this reasonably achievable in the context considered?
- Suppose it is thought likely that about n_1^* cases will become available. Is this sufficient to produce worthwhile conclusions? Would a smaller investigation be appropriate? This is often the more realistic way of studying the appropriate scale of investigation.

- Would it be preferable to stop when a preassigned number of cases has been obtained?

Such issues are commonly formulated in terms of the achievement of a specified power for the significance test of the null hypothesis that $\psi = 0$. However, it may be that many combinations of values for the power and significance level yield the same notional sample size, that determined by the standard error. So, in fact it is simpler, and essentially equivalent, to examine the standard error likely to be achieved. The latter approach focuses on one target quantity, the standard error, which determines the precision of estimation.

For such calculations we need a prior estimate of $\pi_{0.}$, which specifies the marginal distribution of X. This may require a pilot study if there is no previous information available. In many circumstances, however, there is likely to be some existing knowledge about the distribution of X. The highest precision is achieved when $\pi_{0.} = 1/2$, for example when X is formed by the median dichotomy of a rating scale. Then with $c = 1$ the asymptotic variance of $\hat{\psi}$ is $8/n_1$. Thus, to achieve a standard error of 0.2 approximately 200 cases are needed.

If the standard error of the primary contrast of interest is approximately $\sigma/\sqrt{n}$, where n is a measure of the size of the investigation, then, at significance level α, a power of β will be achieved at a value of the true contrast approximately equal to

$$(k_\alpha^* + k_\beta^*)\sigma/\sqrt{n}, \tag{2.33}$$

where k_α^*, say, is the upper α point of the standard normal distribution. More refined calculations are in principle possible in particular cases, but because the specification is typically rather arbitrary such a refinement seems largely unnecessary.

2.6 Logistic dependence

The previous discussion has centred crucially on the equality between the prospective and retrospective log odds ratios. In many contexts it is entirely satisfactory to summarize a contrast by means of the odds ratio or its logarithm. The latter is typically a number not too far from zero in value and hence is conveniently scaled and, in at least some contexts, likely to be relatively stable across replicated studies. Nevertheless it may sometimes be necessary to represent a contrast differently. For example, a *difference* in probabilities may have an important tangible meaning in determining the differences in the numbers of individuals affected in a large population. One possibility is to define

$$\delta = \Pr(X = 1, Y = 1) - \Pr(X = 0, Y = 1) = \pi_{11} - \pi_{01}. \quad (2.34)$$

This in effect determines the difference between the number of cases ($Y = 1$) in a large population with exposures $X = 1$ and $X = 0$. Yet another possibility is to consider the change in the number of cases if a large number of individuals moved from no exposure, $X = 0$, to exposure, $X = 1$, other things being equal; this is determined by

$$\Delta = \Pr(Y = 1 | X = 1) - \Pr(Y = 1 | X = 0). \quad (2.35)$$

Because the differences δ and Δ depend strongly on how frequently the rare outcome $Y = 1$ occurs in the population, they cannot be estimated from the case-control data on their own; in the case-control data, $Y = 1$ has been forced to be relatively frequent. Therefore additional information is essential. One possibility is that $\pi_{.1}$, the overall frequency of $Y = 1$ in the population, can be estimated from independent data,

$$\pi_{.1} = \pi_{01} + \pi_{11}. \quad (2.36)$$

It is easily shown that, for example,

$$\delta = \pi_{.1} \left(\frac{2\pi_{11}}{\pi_{01} + \pi_{11}} - 1 \right) \quad (2.37)$$

and the conditional probability in this expression is directly estimated from the case-control study.

There is a connection between the above point and the discussion in Section 2.4. The essence of the earlier discussion is that the regression coefficient ψ but not the intercept χ in the *sampling model*, that is, the model arising from a case-control study, is correctly estimated by fitting the *inverse model*. We can also make this statement in reverse by saying that in the earlier discussion the regression coefficient ψ but not the intercept λ in the inverse model that would arise in a prospective study, is correctly estimated by fitting the sampling model, that is, by using a case-control study.

Here we establish a connection with the more general account needed to deal with realistic situations. In the general formulation of the inverse model, the binary variable Y defining whether the individual is a case, $Y = 1$, or a control, $Y = 0$, is assumed to have a linear logistic regression on the vector X of explanatory variables, that is,

$$\Pr(Y = 1 | X = x) = L_1(\alpha + \beta^T x), \quad (2.38)$$

Table 2.4 *Summary of data from a case-control study with multilevel explanatory variable.*

	$Y = 0$	$Y = 1$
$X = 0$	r_{00}	r_{01}
$X = 1$	r_{10}	r_{11}
$X = 2$	r_{20}	r_{21}
$\vdots$	$\vdots$	$\vdots$
$X = K$	r_{K0}	r_{K1}

where α, β are unknown parameters representing an intercept and a vector of regression coefficients. For the special case studied in Section 2.4, X is a binary explanatory variable. It will be shown later that we can estimate the regression coefficient β, but not the intercept α, under the sampling model, that is, in a case-control study.

2.7 Explanatory variable with more than two levels

We assumed in much of the discussion above that the explanatory variable or exposure, X, is binary. Extensions to the situation where there are more than two possible values for the discrete variable X are relatively minor. In the most direct approach one possible value of X is taken as a reference level, subsequently denoted by zero. While in a sense the choice of reference level is arbitrary, it is preferable that the zero level should have a special natural meaning as a base for interpretation and that its frequency of occurrence is not small. The notation for data from a case-control study with a multilevel explanatory variable is set out in Table 2.4.

It would be possible to extend the approach based on the generalized hypergeometric distribution, but instead we will assume that the observed frequencies are such that the much simpler large-sample methods may be used. Thus we estimate for each non-zero level of X a log odds ratio

$$\hat{\psi}_x = \log \frac{r_{x1} r_{00}}{r_{x0} r_{01}}, \qquad (2.39)$$

with asymptotic variance

$$1/r_{x1} + 1/r_{00} + 1/r_{x0} + 1/r_{01} \qquad (2.40)$$

and such that, for $x \neq x'$,

$$\text{cov}(\hat{\psi}_x, \hat{\psi}_{x'}) = 1/r_{00} + 1/r_{01}. \tag{2.41}$$

For some purposes, interest may indeed be concentrated on comparisons with the reference level, zero. Then the estimates $\hat{\psi}_x$ and an asymptotic standard error derived from (2.40) summarize the analysis. In other cases different contrasts between the levels of X may be of interest and then the following device is useful.

Write for $x = 0, 1, \ldots, K$

$$\tilde{\psi}_x = \log(r_{x1}/r_{x0}), \tag{2.42}$$

sometimes called a floating estimate. The values of $\hat{\psi}_x$ at two different values of x may be treated as uncorrelated, with asymptotic variances

$$\tilde{v}_x = 1/r_{x1} + 1/r_{x0}. \tag{2.43}$$

It follows that any contrast $\Sigma c_x \psi_x$, with $\Sigma c_x = 0$, is estimated by

$$\sum c_x \tilde{\psi}_x \tag{2.44}$$

with asymptotic variance

$$\sum c_x^2 \tilde{v}_x. \tag{2.45}$$

Note that $\tilde{\psi}_x$ is not to be interpreted on its own, only in the context of contrasts. The gain from considering $\tilde{\psi}$ is in economy of presentation, especially for larger systems. The calculations of precision may be based on the set of variances $\tilde{v}_x$, whereas in terms of the estimates $\hat{\psi}_x$ a full covariance matrix has to be reported unless interest is concentrated entirely on individual comparisons with the reference level.

2.8 Conditional odds ratios: combining estimates

In the above sections we have considered the simplest situation, that of a case-control study with a single binary or categorical exposure. In practice it will almost always be desirable to adjust for other variables. In case-control studies there are two primary motivations for wishing to estimate odds ratios that are conditional on one or more other variables:

- to adjust for confounders, that is, for variables that affect both the exposure X and the outcome Y;
- to adjust for background or intrinsic variables that affect the outcome Y but are not independently associated with the exposure X.

The first situation is illustrated in Figures 3(b), (c) and the second in Figure 3(a).

The ratio of the odds of $Y = 1$ given $X = 1$ and the odds of $Y = 1$ given $X = 0$, conditional on variables W, is

$$\frac{\Pr(Y = 1|X = 1, W = w)/\Pr(Y = 0|X = 1, W = w)}{\Pr(Y = 1|X = 0, W = w)/\Pr(Y = 0|X = 0, W = w)}. \tag{2.46}$$

As in the unconditional situation, this is equal to the ratio of the odds of $X = 1$ given $Y = 1$ and the odds of $X = 1$ given $Y = 0$ conditional on variables W:

$$\frac{\Pr(X = 1|Y = 1, W = w)/\Pr(X = 0|Y = 1, W = w)}{\Pr(X = 1|Y = 0, W = w)/\Pr(X = 0|Y = 0, W = w)}. \tag{2.47}$$

In this section we focus on a categorical variable W or, more generally, on a set of variables W that define two or more strata. In an extension to the earlier notation we let n_{0s}, n_{1s} denote the numbers of cases and controls respectively in stratum s and r_{1s}, r_{0s} denote the numbers of exposed cases and exposed controls respectively in stratum s.

The results of Section 2.3 provide the basis for pooling estimates of log odds ratios ψ from different sources. In many situations it may be reasonable to assume that the association between the exposure and the outcome, in other words the treatment effect, is the same across strata. If this is the case then the data from the different strata can be used to calculate a pooled estimate of the odds ratio of interest. The results leading to this provide a method of testing for the heterogeneity of the treatment effect across values of W, or more generally, across strata.

We begin, as earlier, by considering an exact method of analysis and later outline a simpler, empirical, approach. Under a model in which each stratum has its own nuisance parameters but ψ is the same for all strata, given $r_{.s} = r_{1s} + r_{0s}$ the random variable R_{1s} has a generalized hypergeometric distribution:

$$\Pr(R_{1s} = r_{1s}|R_{.s} = r_{.s}) = \frac{\binom{n_{1s}}{r_{1s}}\binom{n_{0s}}{r_{0s}}e^{r_{1s}\psi}}{\sum_k \binom{n_{1s}}{k}\binom{n_{0s}}{r_{.s} - k}e^{k\psi}}. \tag{2.48}$$

From the combined likelihood for all strata it can be found that the relevant statistic for inference about ψ is $T = \Sigma_s R_{1s}$, that is, the sum of the numbers of exposed cases across strata. The distribution of T is the convolution of the separate hypergeometric or generalized hypergeometric distributions. The main direct usefulness of this is as a test of the null hypothesis, $\psi = 0$, of no difference between the distributions of X in the cases and the controls. Then the exact mean and variance of T are

$$\sum \frac{r_{.s} n_{1s}}{n_{.s}}, \quad \sum \frac{n_{1s} n_{0s} r_{.s} (n_{.s} - r_{.s})}{n_{.s}^2 (n_{.s} - 1)}. \tag{2.49}$$

Especially if the number of strata is not small, under the null hypothesis T will be close to normally distributed and hence a normal-theory test of the null hypothesis $\psi = 0$ can be used, supplemented by a continuity correction because T is an integer-valued random variable.

These procedures are relevant whenever information from several or many 2×2 *contingency tables* are combined. A large number of approximate methods for interval estimation have the aim of emulating the formally exact procedure based on the convolution of generalized hypergeometric distributions.

The maximum likelihood estimate for the common log odds ratio ψ can be found by maximizing the likelihood formed by the product over all strata of the terms (2.48). However, as in the unstratified situation, numerical methods are required. A linear approximation near $\psi = 0$ of the derivative with respect to ψ of the log likelihood arising from (2.48) gives the simple approximation

$$\hat{\psi} = \log \frac{\Sigma r_{1s} (n_{0s} - r_{0s})/(n_{1s} + n_{0s})}{\Sigma r_{0s} (n_{1s} - r_{1s})/(n_{1s} + n_{0s})}. \tag{2.50}$$

When the frequencies are large enough to justify the use of asymptotic methods the analysis becomes more transparent, being based on the calculation and comparison of the mean log odds ratio estimates across strata inversely weighted by variance, that is by the method of weighted least squares. In stratum s the empirical log odds ratio is

$$\hat{\psi}_s = \log \frac{r_{1s} (n_{0s} - r_{0s})}{r_{0s} (n_{1s} - r_{1s})}, \tag{2.51}$$

and its asymptotic variance, denoted v_s, can be found from (2.23). From separate estimates $\hat{\psi}_s$ with asymptotic variances v_s, the pooled estimate is

$$\hat{\psi} = \frac{\Sigma \hat{\psi}_s / v_s}{\Sigma 1 / v_s} \tag{2.52}$$

with asymptotic variance $(\Sigma 1/v_s)^{-1}$. Moreover, the mutual consistency of the different estimates may be tested by the chi-squared statistic

$$\Sigma (\hat{\psi}_s - \hat{\psi})^2 / v_s \tag{2.53}$$

having one fewer degrees of freedom than the number of strata. An extension of the analysis would allow study of the possible dependence of ψ_s on whole-strata explanatory features.

Stratified sampling of cases and, more commonly, controls can be used to ensure that population strata of interest are represented by a sufficient number of individuals. The analyses described in this section could be weakened if only a small number of individuals appeared in some strata. In epidemiological applications sampling strata may be based on sex and age group, for example, giving one stratum for each combination of sex and age group. In other areas of application the strata may be areas of residence, household income brackets or sizes of company. Frequency matching is a special type of stratified sampling in which controls are sampled within strata, on the basis of information about the cases, so as to give a similar distribution of certain features among sampled cases and controls.

The above procedures are also relevant when one is combining results from independent studies that are provisionally assumed to have the same value of ψ. This is covered in further detail in Chapter 10.

2.9 A special feature of odds ratios

In the previous section we outlined methods for combining odds ratio estimates across a range of values of an explanatory variable W, or more generally across strata formed perhaps by combining two or more explanatory variables. It is well known that failure to control for variables that confound the association between an exposure and an outcome will result in biased estimates of the association. However, we have also mentioned the potential for estimating an odds ratio conditional on a background or intrinsic variable which is associated with the outcome Y but not with the exposure X, that is, a variable which is not a confounder, as illustrated in Figure 3(a). We discuss this here in more detail.

If Y is a continuous outcome then in linear situations ignoring W results in no bias in the standard estimate of the effect of X on Y. The reason is that the regression coefficient of Y on X given W is the same as the regression coefficient of Y on X alone when W is connected only to Y, that is, when W and X are marginally independent. For this reason such a conditional mean is described as *collapsible*. There may sometimes be a substantial gain in precision, however, by the inclusion of W in a linear regression model for continuous Y; this gain in precision is due to a reduction in the effective error variance.

The situation is different when Y is a binary outcome and when the association between X and Y is measured by an *odds ratio*. The conditional odds ratio between Y and X given W is not the same as the odds ratio between Y and X ignoring W, even if W is not directly connected to X. Essentially this is so because the combining of frequencies is a linear operation in the data whereas the combining of expected log odds is not; this is referred to as *non-collapsibility*. Non-collapsibility is different from confounding but it has similar implications with regard to controlling for intrinsic variables.

Example 2.2: Non-collapsibility of odds ratios   We will illustrate the above points by considering a numerical example, which follows that given by Gail *et al.* (1984).

Suppose that an intrinsic binary variable W affects a binary outcome Y but is not independently associated with the binary exposure of interest, X. The numbers of individuals in each of the four cells for exposure and outcome are shown in Table 2.5(a) separately for the subgroups with $W = 0, 1$. In this example the marginal distribution of X is the same within the subgroups defined by W. The estimated log odds ratios in the subgroups with $W = 0, 1$ are

$$\hat{\psi}_0 = \log \frac{500 \times 900}{100 \times 500} = \log 9, \quad \hat{\psi}_1 = \log \frac{900 \times 500}{500 \times 100} = \log 9.$$

Because W is not independently associated with X, and W does not modify the association between X and Y, it may seem that we can discount W in the analysis and pool the frequencies in Table 2.5(a) across W. The pooled frequencies are shown in Table 2.5(b). The log odds ratio estimate using the combined frequencies is

$$\hat{\psi}^* = \log \frac{1400 \times 1400}{600 \times 600} = \log \frac{49}{9}.$$

Even though the conditional odds ratio is the same for all values of W, and there is no independent association between X and W, the marginal odds ratio is not equal to the conditional odds ratios. This is what is referred to as the *non-collapsibility* of the odds ratio. The pooled log odds ratio, using the result (2.52), is in fact

$$\hat{\psi} = \frac{\hat{\psi}_0/v_0 + \hat{\psi}_1/v_1}{1/v_0 + 1/v_1} = \log 9,$$

where the v_w are found using (2.23).

Table 2.5 *Example 2.2: non-collapsibility of odds ratios.*

(a) Separately by W.

	$W = 0$		$W = 1$	
	$Y = 0$	$Y = 1$	$Y = 0$	$Y = 1$
$X = 0$	500	500	900	100
$X = 1$	100	900	500	500

(b) Frequencies combined over W.

	$Y = 0$	$Y = 1$
$X = 0$	1400	600
$X = 1$	600	1400

A parameter expressing the dependence of Y on X is said to be collapsible over W if both the following conditions hold:

(1) in the conditional dependence of Y on X given $W = w$ the parameter is the same at all values of w;
(2) in the marginal dependence of Y on X the parameter takes that same value.

Thus, in the example just given, condition (1) is satisfied but (2) is not. Non-collapsibility is in part connected with the nonlinearity of the odds ratio as a function of the data. Note that if

$$E(Y|X = x, W = w) = a(x)b_1(w) + b_2(w)$$

and if $E(b_s(W)|X = x)$ does not depend on x then a generalized form of collapsibility holds with condition (1) above, which requires $b_1(w)$ to be constant. Simple linear regression is a special case and another such occurs when $E(Y|X = x, W = w) = \exp(\beta x + \gamma w)$.

In the situation considered in this section, there appear to be two different odds ratios that might be estimated, one where W is ignored and another where adjustment is made for W, for example by pooling odds ratio estimates from within strata defined by W. The former may be described as the *marginal odds ratio* and the latter as the *conditional odds ratio*.

The implications of this feature of the odds ratio include the following.

- Estimates of marginal odds ratios for X from a case-control study and from a prospective study will differ unless the distribution of W given Y

is the same in the two populations, even if the effect of the exposure on the outcome is the same in the underlying population.

- Estimates of marginal odds ratio for X in case-control studies with different distributions of background variables W will differ even if the effect of the exposure on the outcome is the same in the underlying population.
- Odds ratios for exposures in studies that adjust for different background variables W are not directly comparable.
- The marginal odds ratio from a frequency-matched case-control study will differ from the conditional odds ratio, in which the frequency matching is controlled in the analysis, for example by the pooling of estimates across strata.

In general, therefore, it is likely to be a good idea to estimate odds ratios conditioned on important background or intrinsic variables affecting Y even though they are not confounders.

2.10 Introducing time

We turn now to a situation in which the the population under study is changing over time, as illustrated in Figure 1.5.

The question arises as to how to select controls for cases occurring in a changing population. We begin the discussion by considering a study period $(0, \tau)$ during which the rate at which cases occur, the hazard rate, is constant. For a binary exposure X the hazard rates in the two exposure groups are denoted $h_x, x = 0, 1$. In this situation, where it is constant over a fixed time period time, the hazard rate is often referred to as the *incidence rate* or *incidence density*. If all individuals in the population of interest could be studied, as in a prospective study, then, in the notation of Table 2.2, the hazard rates can be estimated as

$$\hat{h}_0 = (n_1 - r_1)/T_0, \quad \hat{h}_1 = r_1/T_1, \tag{2.54}$$

where n_1 denotes the total number of cases observed, r_1 denotes the number of exposed cases (for which $X = 1$) and T_x denotes the total time at risk of individuals with exposure $X = x$ during the study period $(0, \tau)$.

The *rate ratio* is the ratio of the hazard rates in the two exposure groups, $\kappa = h_1/h_0$ say, and is estimated as

$$\hat{\kappa} = \frac{r_1/(n_1 - r_1)}{T_1/T_0}. \tag{2.55}$$

We now extend to a case-control setting. Suppose that all cases observed in the population during $(0, \tau)$ are included; then $r_1/(n_1 - r_1)$ can be estimated directly. More generally, a random sample of the cases could be used. The ratio T_1/T_0 can be estimated by selecting a sample of control individuals from the population such that the ratio in the sample estimates the ratio in the underlying population. This could be achieved, for example, by selecting control individuals with probabilities proportional to their time at risk during the study period $(0, \tau)$. It follows that, for a constant hazard rate over the study period, the rate ratio can be estimated from a case-control sample. Note that the 'controls' here represent the entire study population at risk of becoming a case during $(0, \tau)$ and therefore the control group may contain some individuals who also appear in the case group. This differs from the situation considered in the earlier sections of this chapter, where the control group was representative of *non-cases* after a period over which cases were ascertained.

In many situations it will be unreasonable to assume a constant hazard rate over a long study period. For example, in an epidemiological context, disease rates typically depend on age. Even if hazard rates change over time, it may often still be reasonable to assume that *ratios* of hazards are constant over time. One simple approach to estimating a rate ratio in this situation is to divide the study period into a series of short consecutive time periods each of length Δ, within which it is reasonable to suppose that (i) the population is unchanging, (ii) the hazard rates within exposure groups are constant.

Extending our earlier notation, we let h_{xk} denote the hazard rate in the kth short time interval for an individual with exposure $X = x$. Assuming the ratio of hazard rates in the two exposure groups to be constant over the whole study period, we have $h_{1k}/h_{0k} = \kappa$ in all intervals k. Again for a moment supposing that the entire study population could be observed, the rate ratio within the kth short time interval can be estimated by

$$\hat{\kappa}_k = \frac{r_{1k}/(n_{1k} - r_{1k})}{(N_{1k}\Delta)/(N_{0k}\Delta)} = \frac{r_{1k}/(n_{1k} - r_{1k})}{N_{1k}/N_{0k}}, \qquad (2.56)$$

where, during the kth short time interval, n_{1k} is the total number of cases, r_{1k} is the number of exposed cases ($X = 1$) and N_{xk} is the total number of individuals in the population at the start of the kth time interval with exposure $X = x$.

In a case-control setting we suppose that one control is selected for each case; the controls are chosen at random from those at risk of becoming a case at the start of the short time interval during which a case occurs. Under this scheme, the *expected* numbers of cases and controls in each of the exposure groups are then as summarized in Table 2.6. This method of

Table 2.6 *Expected numbers of cases and controls selected within the kth short time interval of length* Δ.

	$Y = 0$	$Y = 1$
$X = 0$	$(N_{1k}h_{1k} + N_{0k}h_{0k})N_{0k}/N_{.k}$	$N_{0k}h_{0k}$
$X = 1$	$(N_{1k}h_{1k} + N_{0k}h_{0k})N_{1k}/N_{.k}$	$N_{1k}h_{1k}$

sampling controls is sometimes referred to as *incidence density sampling*. We note that sampling individuals to the case-control study in this way allows for the use of exposures that change over time.

The above procedure for selecting controls is performed within all the short time intervals that make up the study period $(0, \tau)$, giving a series of 2×2 tables, one for each time interval. If the numbers within each table are large enough then the results from the separate tables may be combined using the methods outlined in Section 2.8, subject to an assessment of the consistency of the rate ratio across time intervals. However, if the time intervals are very short, which may be necessary to meet conditions (i) and (ii) above then the data within 2×2 tables may be sparse, making large-sample methods inappropriate.

Another option, which may seem attractive, is to combine the information from all the 2×2 tables into one table, by summing each entry in Table 2.6 over all short time intervals k. Doing this gives for the combined odds ratio from the case-control study, using expected values

$$\frac{\sum N_{1k}h_{1k} / \sum N_{0k}h_{0k}}{\left\{\sum (N_{1k}h_{1k} + N_{0k}h_{0k})N_{0k}/N_{.k}\right\} / \left\{\sum (N_{1k}h_{1k} + N_{0k}h_{0k})N_{1k}/N_{.k}\right\}}.$$
(2.57)

Using the assumption that the rate ratio is the same across the entire study period, $h_{1k}/h_{0k} = \kappa$, the odds ratio in (2.57) is equal to the rate ratio κ only when the ratio of the exposed and unexposed numbers at risk, N_{1k}/N_{0k}, does not change over time. This condition will not be met in an unchanging population or in a population that individuals leave but into which there is no new influx, unless there is no association between exposure and outcome. The condition may be met, however, in a population that people both join and leave, where the changes in the population are such that the ratio of exposed and unexposed individuals is balanced over time.

The odds ratio in (2.57) also estimates κ in the trivial case in which the hazard rate is constant over the entire study period. This combined approach does not accommodate exposures that change over time.

Taking the approach of case-control sampling within short time intervals to an extreme results in the selection of controls for individual cases in continuous time, leading to the possibility that a matched analysis may be the most appropriate under this sampling design. Matched analyses are described in Chapter 3. Incidence density sampling in continuous time is also referred to as *nested case-control sampling*, especially when the underlying population is an assembled cohort. This is discussed in detail in Chapter 7.

2.11 Causality and case-control studies

As with observational studies in general, considerable caution is necessary in giving the conclusions from case-control studies a causal interpretation, even though establishing causality is in some sense a prime objective of many such studies. Indeed, outside the research laboratory it may be fairly rare that any single study, even a randomized experiment, will firmly establish causality.

We shall not discuss these issues in depth in this book. The key principles can best be seen using the following progression: randomized experiment to prospective observational study to retrospective observational study.

In the simplest form of a randomized experiment we have explicitly defined experimental units, such that any two units *might* receive different treatments, that is, exposures, and each unit receives just one treatment. Observations are obtained on each unit, some before randomization and some, called *intermediate*, after randomization but before the final outcome or response variable is defined and measured. In its strong form, the basic principle of analysis in a randomized study comprises the following:

- treatment effects may be adjusted for properties of the units measured before randomization;
- intermediate variables may be analysed in their own right as responses (outcomes) but are otherwise ignored, except in connection with the more detailed task of tracing possible pathways of dependence between treatment and response.

The principle of analysis summarized above is typically achieved by fitting a regression model leading in effect to the estimation of a treatment effect defined conditionally on the background variables. The central idea is that the effect of treatment is defined in terms of changes in response, considered as conditional on the values of the background variables. A separate but important issue is the study of the possible dependence of the exposure effect on background variables, that is, of an exposure by background variable interaction, often better called 'effect modification'. As is consistent with one of the primary principles of experimental design, establishing the absence of significant interactions with background variables is basic to establishing the generalizability of conclusions and, sometimes at least, enhancing the evidence for causality.

A slightly weaker form of the above basic principle is that adjustment may be made for variables measured after randomization but referring to the state of the unit before randomization. Here there is an assumption of no recall bias.

In exceptional cases adjustment may be made for other variables for which there is strong *a priori* evidence that their impact on response is independent of the randomized treatment applied. An objective of the randomization is to convert any impact of unobserved background variables on the response into random error, when assessing the effect of treatments.

In a comparable prospective observational study the key issue is to decide for each exposure (treatment) of interest what features are to be regarded as prior to the risk factor, in terms of the temporal order underlying the data-generating process. These variables *and in general no others* are to be regarded as fixed, as we study hypothetical changes in the risk factor of concern. For the case-control studies of primary interest in this book the same principle holds. In line with the general approach adopted here we consider the comparable prospective studies and notionally hold fixed only those features in a sense prior to the risk factor under consideration. That is, we include only such features in the logistic regression analysis of the case-control status of individuals. Further, in a matched case-control design the matching should be based only on such aspects. The only distinctive aspect that may make case-control studies relatively less secure is that the causal ordering is assessed retrospectively.

In addition to these formal issues, crucial informal considerations apply in assessing causality. R.A. Fisher, when asked how observational studies could be made more indicative of causality, gave the gnomic reply: 'Make your theories elaborate'. This may be interpreted as meaning 'assemble evidence of many different kinds'. Bradford Hill gave a number of conditions, the satisfaction of which made a causal interpretation more likely; he emphasized that they were *considerations* not *criteria*. That is, all might be satisfied and yet causality does not hold, or the converse could happen.

The possibility that an apparently causal effect is totally explained by an unobserved confounder becomes increasingly remote if the effect is large. For, suppose that the apparent effect of a binary exposure X is in fact totally explained by an unobserved binary variable U, causally prior to X; that is, the binary outcome Y is conditionally independent of X given U. We will argue in terms of the risk ratios

$$\mathcal{R}_X = \frac{\Pr(Y = 1 | X = 1)}{\Pr(Y = 1 | X = 0)}, \quad \mathcal{R}_U = \frac{\Pr(Y = 1 | U = 1)}{\Pr(Y = 1 | U = 0)}. \tag{2.58}$$

Because of the assumed conditional independence it can be shown that

$$\mathcal{R}_X = \frac{\mathcal{R}_U \Pr(U = 1 | X = 1) + \Pr(U = 0 | X = 1)}{\mathcal{R}_U \Pr(U = 1 | X = 0) + \Pr(U = 0 | X = 0)}. \tag{2.59}$$

Therefore

$$\mathcal{R}_X \leq \mathcal{R}_U, \quad \mathcal{R}_X \leq \frac{\Pr(U = 1 | X = 1)}{\Pr(U = 1 | X = 0)}. \tag{2.60}$$

These inequalities, called Cornfield's inequalities, have, in a particular example, the following interpretation. The risk of lung cancer for smokers relative to non-smokers is roughly 10. For this to be solely explained by an unobserved confounder it is necessary first that the unobserved confounder itself has an associated risk ratio of at least 10. Second, the frequency of the unobserved confounder among the smokers has to be at least 10 times that among non-smokers. Thus it is in this context very hard to explain the large estimated risk ratios through an unknown unobserved confounder. This conclusion does not apply to an apparently large *difference* in rates of occurrence.

Notes

Sections 2.1 and 2.2. Various other names have been proposed for the log odds. Barnard (1949) proposed 'lods' and regarded lods as in some sense more basic than probability itself. A very large number of measures have been proposed for assessing dependence in a 2×2 contingency table. For a review, see Goodman and Kruskal (1954, 1959, 1963). Edwards (1963) showed that if a measure of dependence in a 2×2 table is to be essentially unchanged by the interchange of rows and columns it must be the odds ratio or an equivalent. This is highly relevant to the study of case-control studies, although of course it does not imply that the odds ratio is the best basis for interpretation.

Section 2.3. The reduction of the data to the cell frequencies shown in Table 2.2 is formally justified by sufficiency, which in turn depends on the assumed independence of the individuals and the strong homogeneity assumptions made. The use of the generalized hypergeometric distribution as the basis for inference for the odds ratio is due to Fisher (1935). A seemingly never-ending controversy surrounds the test of significance, arising almost entirely from the requirement in the formal Neyman–Pearson theory of exactly achieving artificially prespecified levels of probability rather than of using tail areas as a limited measure of evidence. In the binomial distribution written in exponential family form, the canonical parameter is the log odds, $\log\{\theta/(1 - \theta)\}$; the odds ratio ψ emerges as the parameter in the resulting conditional test statistic distribution essentially because it is a difference of canonical parameters. That this is also the appropriate parameter for interpretation in a case-control context appears as a happy chance, but whether chance happenings occur in mathematical discussions may be doubted.

Section 2.3.2. Approximate methods run into serious problems with small counts and obviously worse than serious problems if zero frequencies are encountered. Some improvement is obtained by adding $1/2$ to all the frequencies in (2.18) and (2.23). For a theoretical justification see Cox (1955). The use of the Poisson representation stems from Fisher (1922). Its use is a convenient computational device; all the relevant formulae could be derived, but more clumsily, from the covariance matrix of the multinomial distribution. For a more detailed discussion see Bishop *et al.* (1975). If a variable R has a binomial distribution corresponding to n trials, random variables such as the log odds, $\log\{R/(n - R)\}$, are improper in the sense that, no matter how large n is, there is a non-zero probability that $R = 0$ or $R = n$. Statements about the distribution of the log odds thus have either to be made conditionally on $R \neq 0, n$ or regarded as asymptotic statements about the distribution function of the random variable as n increases; the latter route is taken here.

Section 2.6. The choice of the scale on which effects are to be defined arises in many contexts. In medical statistics and epidemiology it is frequently a choice between ratios of probabilities (or odds) and differences of probabilities. While the choice is often made, as here, essentially on the grounds of statistical convenience or necessity, it is in principle a subject-matter issue. That odds ratios tend to be numbers fairly close to unity has a mnemonic advantage and may mean that they are relatively stable across similar studies. Differences in probabilities are often small numbers, and so not likely to be stable, but they have the major advantage that they are easily related to the numbers of individuals implicitly likely to be affected by potential changes in exposure. The estimation of quantities other than the odds ratio in a case-control study was discussed by Benichou and Gail (1995) in the context of a population-based case-control study and in more general terms by King and Zeng (2002).

Section 2.7. The use of floating estimates stems from an older practice in the analysis of variance in which conclusions were reported by tables of two-way and higher-way tables of marginal means rather by differences from a baseline level. In more complex cases the simplified representation is only an approximation. See Ridout (1989), Easton *et al.* (1991) and Firth and de Menezes (2004).

Section 2.8. See Cox (1958a) for an outline theoretical justification of the methods outlined in this section. Mantel and Haenszel (1959) also discussed the combination of information from 2×2 tables, and the simplified pooled estimate in (2.50) is often referred to as the Mantel–Haenszel estimate. Woolf (1955) gave the weighted least squares estimates. Cox and Snell (1989, Section 2.5) summarized some of the many methods suggested for estimation of the common log odds ratio. See also Breslow (1981) for a summary of pooled estimators, and, for a discussion of variance estimators, Breslow and Liang (1982), Phillips and Holland (1986), Robins *et al.* (1986) and Mantel and Hauck (1986). For a test of homogeneity, see Breslow and Day (1980, pp. 142–3).

Section 2.9. The non-collapsibility of odds ratios is discussed by, for example, Gail *et al.* (1984) and Robinson and Jewell (1991).

Section 2.10. The estimation of rate ratios using incidence density sampling in case-control studies was described by Miettinen (1976). See also Greenland and Thomas (1982). A useful overview of different methods of control sampling in case-control studies was given by Rodrigues and Kirkwood (1990). Breslow and Day (1980) (pp. 72–3) briefly described how long study periods could be divided into smaller intervals and the resulting estimates combined.

Section 2.11. Causality has implicitly been the objective of statistical study for a long time (Bradford Hill, 1965, Cornfield *et al.*, 1959, Cochran, 1965). Recent intensive interest in the subject has tended to focus on formal representations; see the wide-ranging collection of papers in Berzuini *et al.* (2012). For a somewhat contrasting discussion see Cox and Wermuth (2004). Bradford Hill's considerations or guidelines are discussed in many books on epidemiology; for the original, see Bradford Hill (1965). The most controversial of these guidelines is that effects are more likely to be found causal if they are specific. That is, a risk factor that appears to have an adverse effect on many distinct outcomes is thereby less likely to be causal. For a striking example of the detailed discussion desirable to establish causality, see the early discussion of smoking and lung cancer (Cornfield *et al.*, 1959, reprinted in 2009). The inequalities discussed at the end of this section form an appendix to Cornfield's paper. For a general discussion of observational studies, see Cochran (1965), where the comment of R.A. Fisher is recorded. For a systematic account of graphical models to study the relationships that can be estimated from case-control studies, including causal relationships, see Didelez *et al.* (2010).

3

Matched case-control studies

- The individual matching of controls to cases in a case-control study may be used to control for confounding or background variables at the design stage of the study.
- Matching in a case-control study has some parallels with pair matching in experimental designs. It uses one of the most basic methods of error control, comparing like with like.
- The simplest type of matched case-control study takes a matched-pair form, in which each matched set comprises one case and one control.
- Matched case-control studies require a special form of analysis. The most common approach is to allow arbitrary variations between matched sets and to employ a conditional logistic regression analysis.
- An alternative analysis suitable in some situations uses a regression formulation based on the matching variables.

3.1 Preliminaries

An important and quite often fruitful principle in investigating the design and analysis of an observational study is to consider what would be appropriate for a comparable randomized experiment. What steps would be taken in such an experiment to achieve secure and precise conclusions? To what extent can these steps be followed in the observational context and what can be done to limit the loss of security of interpretation inherent in most observational situations?

In Chapter 2 we studied the dependence of a binary outcome, Y, on a single binary explanatory variable, the exposure, X. The corresponding experimental design is a completely randomized comparison of two treatments. One widely used approach to enhancing the precision of such an experiment is to employ a matched-pair design, in which the experimental units are arranged in pairs in such a way that there is reason to expect very similar outcomes from the two units in a pair were they to receive the same treatment. The randomization of treatments is then constrained in such a way that the two units forming a pair get different treatments. In a case-control study a design aspect to be decided by the investigator is the choice of controls. While the analysis and purpose of this aspect is rather different for continuous and for binary variables, there are a number of broad parallels.

In the discussion in Chapter 2 it was in effect assumed that the controls are a random sample from the target population formed from all possible controls. We also extended the discussion to stratified random sampling and the special case of frequency matching. We now suppose that controls are chosen case by case, each control matching the corresponding case on selected variables. This raises the following three issues.

- Precisely how is this matching to be achieved?
- What are the implications of the matching for the analysis of the resulting data?
- What are the implications for interpretation?

In a slightly different terminology, matching is a type of conditioning. By comparing individuals matched with respect to certain features we are, in effect, conditioning our exposure comparison on those features. It is essential that this conditioning is appropriate for the issue under analysis.

Again we begin by considering the simplest situation, that of a binary outcome Y and a binary exposure X. The variables on which the case-control matching is based are denoted by M. In general, for each case one or more matched controls may be chosen or, indeed, matched sets may comprise multiple cases and controls, though this last situation is less usual. In order to set out an approach to analysing matched case-control data, this chapter focuses primarily on the simplest situation, in which one control is matched to each case, otherwise known as 1 : 1 matching or pair matching. Some generalizations are discussed in the later sections.

3.2 Choice of matched controls

In an experimental design the choice of matching units is made *before* the random allocation of treatments. It is thus impossible for the criterion used for matching to be influenced by the treatment applied to a particular unit. The treatment comparison is conditional on all features referring to the units before randomization and therefore, in particular, on the features determining matching.

Time

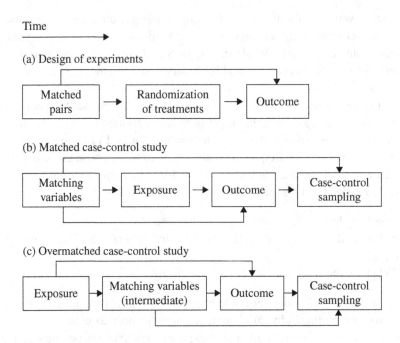

Figure 3.1 Sequence of dependence. (a) Matching in a randomized experiment. (b) A matched choice of controls in a case-control study. (c) Overmatched controls.

The corresponding condition in a case-control study is that the variables M used as a basis for matching must be causally prior to the explanatory variable X under study. That is, the object of interest is in principle the conditional dependence of the outcome Y on X given M. Typical examples of matching variables are thus intrinsic features of the study individuals such as sex, age and, depending on the context, socio-economic class.

It is common to match on well-known confounders which are prior to the explanatory variable of primary concern but which are not themselves of primary interest. Matching has also sometimes been used primarily just as a convenient method for selecting a control group. If the matching variables are not related to the outcome then some insensitivity may result. Matching potentially allows several variables to be adjusted for at once, whereas an adjustment for several variables at once in an unmatched study might quickly become difficult using the stratified methods discussed in Chapter 2.

Matching on given variables effectively eliminates the possibility of investigating the independent effects of those variables on the outcome. Hence care should be taken not to eliminate explanatory variables of potential interest in this way. We leave a more detailed discussion of the relative merits of unmatched and matched case-control studies to Chapter 4.

In implementing matching there may be some choice between insisting on an exact match with respect to one or a small number of features and forcing an approximate match with respect to a larger number of properties. The pool of available controls may not always be large enough to insist on exact matching on several variables, and this would result in a loss of some cases from the study if a match cannot be found. To avoid the loss of many cases from the study, the matching criteria may therefore sometimes need to be relaxed. For example, if age to the nearest year is used as a matching variable then for some cases it may not be possible to find a control of the same age. Hence one may have to relax the matching criterion and select controls whose age is within, say, three years of that of the case. Where the matching is quite crude, it may be necessary to fit the data to an appropriate model in order to account for residual dependence.

Example 3.1: A matched study A case-control sample within the EPIC-Norfolk cohort was used to investigate the association between dietary fibre intake, measured using seven-day diet diaries, and the risk of colorectal cancer. This was part of a larger collaborative study, the UK Dietary Cohort Consortium, and the study results were given by Dahm *et al.* (2010). The cases were all men and women diagnosed with colorectal cancer over the period from their entry to the underlying prospective cohort study, between 1993 and 1998, until the end of 2006. The diet diaries were completed at entry to the cohort study and, to avoid the possibility of bias from individuals who changed their diet because of factors arising from undiagnosed colorectal cancer, cases diagnosed within one year of completion of their diary were excluded. Cases with a prior cancer were also excluded. Each case was matched to four controls, on sex, age at entry to the study (within three years) and date of diary completion (within three months). The controls were also required to have at least as much follow-up as the case, that is, to have survived as long as the case in the study without death or loss to follow-up. This 'time matching' was desirable because of the rather long follow-up time for the underlying cohort. The study comprised 318 colorectal cancer cases and 1272 controls.

3.3 Analysis of matched data

Suppose that data are available on n matched pairs each formed from one case, $Y = 1$, and one control, $Y = 0$, matched with respect to variables M. The explanatory variable of interest X is observed for all selected

Table 3.1 *Possible values of the binary explanatory variable X within a case-control pair in a pair-matched case-control study.*

	Case	Control
Concordant	0	0
Discordant	1	0
Discordant	0	1
Concordant	1	1

individuals. We assume for the moment that X is binary. The pair of binary exposures from each case-control pair can take one of four possible forms, which are summarized in Table 3.1.

Simple comparisons of the distributions of the exposure in the case and control groups are possible. However, we focus on using the odds ratio as a measure of association between exposure and outcome. There are broadly three approaches that might be used for the analysis of matched case-control data.

(1) Ignore the matching and analyse the data as if the controls had been chosen at random. We return to this in Section 4.4.

The remaining two approaches retain the pairing of the cases and controls, as is typical.

(2) In the second and most commonly used method, we assume in effect that the pairs differ arbitrarily but that each pair allows the estimation, as before, of the log odds ratio of interest; this is assumed to be the same for all pairs, conditionally on the features defining the matching.
(3) In the third approach, which also retains the matching information, at least in part, the variation between pairs is described in some simple form, for instance by a model with a limited number of unknown parameters representing, for example, regression on the particular variables M used to define the matching. This last approach assumes more but, by being more economical in parameters, may sometimes lead to higher real or apparent precision in the estimated contrast of interest. This type of analysis is discussed further in Section 3.5.2.

We consider first approach (2), in which we use a representation of arbitrary variation between pairs. This is the method normally recommended for such data, under the name *conditional logistic regression*. In matched pair u for $u = 1, \ldots, n$, we consider probabilities generalizing those of Chapter 2, writing for the variables associated with that pair

$$\pi_{u;xy} = \Pr(X_u = x, Y_u = y), \tag{3.1}$$

where the exposure X_u and outcome Y_u refer to members of the population generating pair u. A table similar to Table 2.1 could therefore be formed for each of the two subpopulations generating the pair u.

Exactly as before, we have a population model, a sampling model and an inverse model used for interpretation. The difference is that there is now one such model relating to each matched pair. The odds ratio relating to pair u is

$$(\pi_{u;11}\pi_{u;00})/(\pi_{u;10}\pi_{u;01}). \tag{3.2}$$

Only one matched pair gives information about this odds ratio. Clearly there is far too little information in the data to proceed without simplification. The most direct approach, in the absence of additional information, is to assume that the log odds ratio ψ is the same for all pairs, that is, that

$$(\pi_{u;11}\pi_{u;00})/(\pi_{u;10}\pi_{u;01}) = e^{\psi} \tag{3.3}$$

for $u = 1, \ldots, n$.

We let $X_{u;0}$ denote the value of the binary explanatory variable X for the control in the uth matched pair, and $X_{u;1}$ denote the value of X for the case in the uth matched pair. The odds ratio in (3.3) can be expressed differently. We write, for the sampling model, that for the control in the uth pair, that is, the member with $Y_u = 0$,

$$\Pr(X_{u;0} = 1) = L_1(\chi_u),$$

whereas for the case in the uth pair, $Y_u = 1$, we have, say,

$$\Pr(X_{u;1} = 1) = L_1(\psi + \chi_u).$$

There are four possible outcomes $(x_{u;1}, x_{u;0})$ for pair u, namely $(0, 0)$, $(1, 0)$, $(0, 1)$, $(1, 1)$, as shown in Table 3.1. The likelihood derived from pair u in the sampling model is

$$L_1(\chi_u)^{x_{u;0}} L_0(\chi_u)^{1-x_{u;0}} L_1(\chi_u + \psi)^{x_{u;1}} L_1(\chi_u + \psi)^{1-x_{u;1}}, \tag{3.4}$$

where $L_0(x) = 1 - L_1(x)$. This likelihood can be written as

$$\frac{e^{\chi_u x_{u;0}}}{(1 + e^{\chi_u})} \frac{e^{(\chi_u + \psi)x_{u;1}}}{(1 + e^{\chi_u + \psi})}. \tag{3.5}$$

The full likelihood is formed from this by taking the product over u values and is therefore

$$\frac{\exp(\Sigma \chi_u x_{u;0}) \exp\{\Sigma(\chi_u + \psi)x_{u;1}\}}{\Pi(1 + e^{\chi_u})\ \ \Pi(1 + e^{\chi_u + \psi})}. \tag{3.6}$$

The contribution to the likelihood in (3.5) is the only part of the total likelihood involving χ_u; that is, the only information we have about χ_u is that which comes from the case and control in the uth matched pair. Because of the very small amount of information available about the parameters χ_u it would be a mistake to estimate ψ by maximizing the above full likelihood. The consequences of doing this are discussed in Section 3.7.

A different form of analysis is therefore required. For this, note that it follows from (3.5) that $x_{u;.} = x_{u;0} + x_{u;1}$ is the minimal sufficient statistic for χ_u. Therefore the only way to achieve a confidence interval or statistical significance test with coverage properties that do not depend on χ_u is by using the conditional distribution of the data in the pair, given the pair total $x_{u;.}$. That is, we ask: what is the probability that we observed $x_{u;1} = k$ and $x_{u;0} = l$ given that $x_{u;1} + x_{u;0} = k + l$? Now, this conditioning statistic $x_{u;.}$ takes values $(0, 1, 1, 2)$ corresponding to the four possible outcomes for pair u, as listed in Table 3.1. In the so-called concordant pairs, $(0, 0)$ and $(1, 1)$, the pair of values given the conditioning statistic is completely determined and therefore provides no information about ψ. That is, we have

$$\Pr(X_{u;1} = 0 | X_{u;0} + X_{u;1} = 0) = 1,$$
$$\Pr(X_{u;1} = 1 | X_{u;0} + X_{u;1} = 2) = 1.$$

It is only the discordant pairs, namely those in which case and control yield different values of X, that such information is available. In fact

$$\Pr(X_{u;1} = 1 | X_{u;0} + X_{u;1} = 1) = L_1(\psi). \tag{3.7}$$

It follows that inference about ψ, treating $\chi_1, \ldots, \chi_n$ as unknown nuisance parameters, is obtained as follows. First, we let n_{ij} denote the number of case-control pairs in which the case has exposure $X = i$ and the control has exposure $X = j$ ($i, j \in \{0, 1\}$), that is, $(x_{u;1}, x_{u;0}) = (i, j)$. The matched-pair case-control data can therefore be summarized as in Table 3.2. The numbers of discordant pairs $(0, 1)$ and $(1, 0)$ are n_{01} and n_{10}, respectively. Thus n_{10} is the observed value of a random variable N_{10} having, conditionally on the total number $n_D = n_{10} + n_{01}$ of discordant pairs, a binomial distribution with parameter $L_1(\psi)$ and index n_D. The

Table 3.2 *Number of pairs of each type in a matched-pair case-control study with binary explanatory variable X.*

	Case	
Control	$X = 0$	$X = 1$
$X = 0$	n_{00}	n_{10}
$X = 1$	n_{01}	n_{11}

conditional likelihood from which to make inferences about ψ has the binomial form

$$\Pr(N_{10} = n_{10} | N_{10} + N_{01} = n_{\mathrm{D}}) = \binom{n_{\mathrm{D}}}{n_{10}} L_1(\psi)^{n_{10}} L_0(\psi)^{n_{01}}. \qquad (3.8)$$

Methods of inference for a binomial parameter that do not require large-sample theory could now be used. Instead, we will follow the approach used for more complicated problems, which is based on maximization of the likelihood with respect to ψ.

Such discussion centres on the calculation of the likelihood function of the data or sometimes of a modification of that likelihood. This is followed by maximization to produce an estimate and its large-sample standard error, from which approximate confidence intervals can be calculated. For a quick review of these procedures, see the appendix.

This approach gives the estimated log odds ratio as

$$\hat{\psi} = \log \frac{n_{10}}{n_{01}}. \qquad (3.9)$$

From the inverse of the observed information, or by the arguments of Section 2.3.2, the large-sample variance of the log odds ratio estimate $\hat{\psi}$ is

$$\frac{1}{n_{10}} + \frac{1}{n_{01}}. \qquad (3.10)$$

Example 3.1 continued We will use data from the study described in Example 3.1 to illustrate the methods outlined above. The study comprised 318 colorectal cancer cases with four controls matched to each case. In this example we consider just the first control in each matched set, which results in a pair-matched case-control study. The exposure of interest in this illustration is a binary variable indicating can average daily fibre intake of < 15 grams/day or ≥ 15 grams/day. The data are summarized in Table 3.3.

Table 3.3 *The number of pairs of each type in a
pair-matched study with a binary explanatory variable
indicating a fibre intake of < 15 or ≥ 15 grams/day in a
study of fibre intake and colorectal cancer risk.*

| | Cases | |
Controls	< 15 g/day	≥ 15 g/day
< 15 g/day	116	70
≥ 15 g/day	83	49

The estimated log odds ratio is

$$\hat{\psi} = \log \frac{70}{83} = \log 0.843 = -0.170,$$

with asymptotic variance $1/70 + 1/83 = 0.026$ and hence standard error
0.162.

A 0.95-level confidence interval for ψ is therefore $(-0.488, 0.148)$. This
pair-matched study did not provide clear evidence of a reduced risk of
colorectal cancer among individuals with fibre intake of ≥ 15 grams/day
compared with those with a lower intake.

3.4 Discussion

The possibly unexpected form of analysis outlined in the previous section
needs qualitative discussion. Consider two extreme situations. In the first
there are 1000 case-control pairs $(0, 0)$ and 1000 pairs $(1, 1)$, and no others.
The second situation is the same except for the addition of 10 pairs $(1, 0)$,
that is, 10 pairs in which $X = 1$ for the case and $X = 0$ for the control,
and no pairs $(0, 1)$. The data are displayed in Table 3.4. Then the analysis
just developed leads to the conclusion that there is no evidence about the
comparison of cases and controls for the former situation whereas in the
latter the extreme split of the 10 discordant cases is very strong evidence that
cases are relatively more likely to have $X = 1$ than are controls. Yet an initial
assessment might well, especially for the former situation, be that there is
strong evidence that any difference between cases and controls is very small.

The requirement that the inference about the log odds ratio ψ does not
depend on the nuisance parameters χ_u, which, in theory at least, may take
any real values, is strong, in a sense often too strong. If, for example, the

Table 3.4 *Number of pairs of each type in pair-matched studies in two situations.*

(a) Situation 1.

	Cases	
Controls	$X = 0$	$X = 1$
$X = 0$	1000	0
$X = 1$	0	1000

(b) Situation 2.

	Cases	
Controls	$X = 0$	$X = 1$
$X = 0$	1000	10
$X = 1$	0	1000

1000 values of χ_u associated with the cases with $X = 1$ happen to be very large and positive and the 1000 values associated with the cases with $X = 0$ are very large and negative then, whatever modestly large value of ψ might hold, the first set of pairs is virtually certain to yield the values $(1, 1)$ and the second set the values $(0, 0)$. That this situation holds is totally consistent with the data and with the initial specification, because it places no constraints on the possible values of the χ_u. That is why, under this specification, the 2000 concordant pairs are genuinely uninformative about ψ and why only the 10 discordant pairs are judged relevant. If the χ_u are in any reasonable sense stochastic then the split of values just mentioned suggests the presence of an unobserved risk factor highly predictive of both X and Y, and the search for this risk factor or factors should become a research objective. Note that also, in view of the method of sampling, the frequencies give no information about the frequency of cases in the population.

From a general perspective it may be argued that a model with one parameter of interest and a large number n of nuisance parameters has too many parameters in the light of the amount of information in the data. Thus, while inference about ψ is technically possible via the conditional analysis, the high degree of caution in the specification comes at a cost that is considerable and indeed possibly sometimes too high. There are two possible ways of resolving this dilemma, which we consider in the next section.

3.5 Alternative analyses of matched data

3.5.1 A random effects analysis

We now consider different approaches to the analysis of matched data. There are two issues to consider. One concerns the precision achieved. The other concerns the comparability of case versus control contrasts as assessed in different models. The second point will be largely deferred to Chapter 4. The log odds conditioned on the variables defining the matching is not the same as the log odds from the two-way table of overall frequencies, which is a log odds marginalizing over those variables.

In the first approach we suppose that the nuisance parameters have a frequency distribution that can be represented as follows. We assume that the random variables $\chi_1, \ldots, \chi_n$ are independent and identically distributed with, say, a normal distribution of mean μ_χ and variance σ_χ^2. Such a representation, at least provided σ_χ is fairly small, effectively excludes extreme values of χ. Under this model the multinomial frequencies of the four possible case-control exposure patterns, $(0, 0), \ldots, (1, 1)$, are defined by the three parameters $\psi, \mu_\chi, \sigma_\chi$. No test of consistency of data and model is possible, at least with the information specified here, and moreover the amount of information about σ_χ may be very small, in which case it may be preferable to specify a value of σ_χ *a priori* and to do a sensitivity analysis on this choice. In fact this approach, while in some ways natural and appealing, is rarely if ever used. It can be shown formally, and is clear on general grounds, that if this random effects formulation is adopted with an essentially non-parametric rich family of distributions for the χ_u then the analysis becomes in fact equivalent to that for the conditional approach.

3.5.2 An adjustment analysis

In another approach we represent the inter-pair variation by a linear regression. Let m_u denote the values of M in the uth matched set. Various regression-like representations are possible, depending on the nature of the matching variables; for example one can take the χ_u to have a normal-theory linear regression on the vector m_u. The resulting log likelihood then involves ψ, and the parameters defining the linear regression and maximum likelihood can be applied.

A simple and often quite realistic situation occurs when the matching variables in effect define a small number q of distinct sets, for example males below 50 years or females below 50 years. Provided there are a reasonable number of individuals in each set, an efficient analysis is based on forming q 2×2 tables corresponding to the unconditional analyses of the separate sets. The formally exact analysis of each table refers the number of individuals with $(X = 1, Y = 1)$ to a generalized hypergeometric distribution, and the combined analysis uses the overall total of such individuals referred in principle to a convolution of generalized hypergeometric distributions. See, in particular, the discussion in Section 2.8.

If the frequencies involved are not small then the large-sample approach described in Section 2.8 is to be preferred.

There are some types of matching criteria that are not easily summarized in one or more regression variables. For example, if cases are matched on 'area of residence' then it may be difficult to capture everything about this in regression variables, though attempts could be made to at least partially capture the matching variable, for example using the available measures of social deprivation within geographical areas. If the matching cannot be completely captured in an unconditional analysis, using a stratified analysis or a regression approach, then the resulting odds ratio estimates may be biased. However, if the variables on which the matching is based can be easily captured then there may be some gain in efficiency by adopting an unmatched analysis that accounts for the matching variables in unconditional stratified or regression analyses.

There are some specific situations in which an unconditional analysis of matched case-control data may be necessary, for example if there are missing data in the explanatory variables in some matched controls or if it is of interest to investigate the association between the explanatory variable and the outcome within subgroups of a non-matching variable. As noted earlier, matching based on a given variable excludes the possibility of assessing the independent association of that variable with the outcome in a conditional analysis. These issues are discussed further in Chapter 4.

3.6 Overmatching

There are circumstances in which an incorrect choice of matching variables could result in a biased estimation of the association between the main exposure of interest and the outcome. In particular, if we were to match cases with controls using a variable potentially affected by the feature X then we would be addressing a different question. Thus if we were studying the incidence of cardiovascular events Y in its dependence on a life-style feature X and we used in-study blood pressure as a matching variable M then we would be assessing the effect on Y of X given M. If the effect of X were to be solely to influence blood pressure and thence Y then a null effect of X would be found after matching on M. The primary aim in such an example would, in fact, typically be to study the whole impact on Y of X, and a matching on blood pressure would be misleading. Such a use of matching, inappropriate in almost all contexts, may be called *overmatching*. See Figure 3.1.

Another type of overmatching occurs when the matching variable is effectively measuring the same underlying feature as is being measured by the main exposure of interest. Matching on such a variable will clearly result in a distorted, and likely close to null, observed association between the main exposure and the outcome. A related situation occurs where the matching is based on a background variable that is correlated with the exposure but not with the outcome. This will result in reduced precision, as the correlation of the matching variable with the exposure will give rise to more concordant sets.

Example 3.2: An example of overmatching   Marsh *et al.* (2002) presented an example of overmatching of the second type in a case-control study of occupational exposure to radiation and leukaemia mortality. The case-control sample was part of a cohort of workers who had been followed up for a number of years after recruitment. The specific exposure was the cumulative radiation dose and was measured from information recorded on film badges worn by workers as a routine.

This study comprised 37 cases who were workers at one particular nu-clear processing plant and had died from leukaemia. Two different sets of matched controls were investigated, with differing matching variables. The first set of matching variables were site, sex, involvement in the same type of work (office work or work involving the handling of radioactive mate-rial), date of birth within two years and date of entry to the cohort within two years. The controls were also required to be alive at the corresponding case's date of death. The second set of matching variables was the same except for the exclusion of matching on date of entry to the cohort. The two sets of controls were therefore described as 'fully matched' and 'partially matched' respectively. In both situations, each case was matched to four controls.

The fully matched case-control study gave an estimated log odds ratio per unit increase in cumulative radiation dose (measured in sieverts) of -0.404 with standard error 0.865. The log odds ratio estimate from the partially matched study was 1.565 with standard error 1.035. Although neither result gave strong evidence of an association, that from the fully matched study contradicted evidence from studies of the full cohort, which found evidence of increased risk of leukaemia mortality with increasing cumulative radiation dose. The result from the partially matched study, however, was supportive of the cohort study results.

The authors suggested that the discrepancy between the results from the fully matched and from the partially matched studies was due to a type of overmatching in the fully matched study, arising from the additional matching on time of entry to the cohort. The reason for including this as a matching variable had been that leukaemia risk is thought to differ over calendar time. However, further investigations showed that the occupational radiation dose was also dependent on calendar time. Hence, the exposure, that is, the cumulative radiation dose, was determined partly by date of entry to the study. The authors concluded that 'As well as eliminating the effect of calendar time, this seems to have had the effect that workers in the same matched set have broadly similar recorded doses. The apparent overmatching on date of entry has distorted the parameter estimate of the risk of leukaemia on cumulative dose by introducing matching (at least partially) on dose.'

3.7 Further theoretical points

Finally, we turn again to the decision to conduct a *conditional* analysis of matched case-control data. There may be a temptation to deal with the matched-pair analysis by a direct use of maximum likelihood, obtaining an estimate by maximizing the full unconditional likelihood (3.6) with respect to $(\psi, \chi_1, \ldots, \chi_n)$. The concordant case-control pairs make no contribution to the estimate of ψ from the unconditional likelihood, and it can be shown that, for one-to-one matching, the profile unconditional likelihood for ψ is

$$
n_{10} \left\{ \frac{\psi}{2} - \log\left(1 + e^{-\psi/2}\right) - \log\left(1 + e^{\psi/2}\right) \right\}
$$
$$
+ n_{01} \left\{ -\frac{\psi}{2} - \log\left(1 + e^{-\psi/2}\right) - \log\left(1 + e^{\psi/2}\right) \right\}. \quad (3.11)
$$

Differentiation with respect to ψ gives the estimated log odds ratio as

$$
2 \log \frac{n_{10}}{n_{01}}, \quad (3.12)
$$

which is twice the consistent log odds ratio estimate from the conditional analysis, given in (3.9). It can also be shown that the asymptotic variance of the unconditional log odds ratio estimate in (3.12) is twice that of the conditional log odds ratio estimate.

Although the unconditional analysis using the full likelihood in (3.6) might seem a good way of analysing the matched case-control data, it results in an inconsistent estimate of the odds ratio, the relative magnitude of the inconsistency becoming more severe as the true odds ratio increases. The reason is that maximum likelihood theory assumes that the parameter space is fixed, while the number of observations, matched sets in this case, tends to infinity. In the unconditional analysis, however, the dimension of the parameter space increases as the number of observations increases.

3.8 Generalizations: several controls per case

The most immediate generalization of the previous discussion involves c matched controls per case and is referred to as $c : 1$ matching. Studies in which the pool of potential controls is not large, and perhaps in which the matching criteria are stringent, may often result in a variable number of controls matched to each case. Such studies would typically aim to obtain a chosen number of controls per case, for example five, but it would be accepted that some cases may have fewer than the target number of matched controls.

For extra generality here, suppose that in the uth matched set there are c_{u0} controls and c_{u1} cases. The value of the exposure X is obtained for each individual; the formal logistic sampling model remains the same as in Section 3.3. In an extension of the previous notation we let $X_{u;0k}$ denote the exposure status of the kth control in the uth matched case-control set and $X_{u;1l}$ denote the exposure status of the lth case in the uth matched case-control set. In the uth matched set let d_{u1} denote the observed number of cases with $X = 1$, and let d_{u0} denote the observed number of controls with $X = 1$. Inference for ψ may then be made, conditionally on $d_{u.} = d_{u0} + d_{u1}$, by finding the conditional probabilities of observing d_{u1} exposed cases and d_{u0} exposed controls in the uth matched set,

$$\Pr\left(\sum_{l=1}^{c_{u1}} X_{u;1l} = d_{u1} \,\middle|\, \sum_{l=1}^{c_{u1}} X_{u;1l} + \sum_{k=1}^{c_{u0}} X_{u;0k} = d_{u.}\right).$$

The contribution to the conditional likelihood from set u is

$$\frac{\binom{c_{u1}}{d_{u1}}\binom{c_{u0}}{d_{u0}} e^{d_{u1}\psi}}{\sum_{k} e^{k\psi}\binom{c_{u0}}{d_{u.} - k}\binom{c_{u1}}{k}}, \tag{3.13}$$

where the sum in the denominator is over k from 0 to $d_{u.}$. There is effectively no contribution to the conditional likelihood from matched sets in which all cases and all controls give the same value of X, that is, if a matched set is concordant.

We find that in the case of a pair-matched case-control study the log odds ratio estimate (3.9) and its asymptotic variance (3.10) take simple forms. There is a closed form solution for the log odds ratio estimates for two controls per case, given by

$$\hat{\psi} = \log\left\{Q + Q^2 + \left(\frac{n_{10} + n_{11}}{n_{01} + n_{02}}\right)^{1/2}\right\}, \qquad (3.14)$$

where n_{ij} denotes the number of matched sets where the case has exposure $X = i$ and where j controls have exposure $X = 1$, and

$$Q = \frac{4(n_{10} - n_{02}) + n_{11} - n_{01}}{4(n_{10} + n_{02})}$$

with the asymptotic variance of $\hat{\psi}$

$$e^{-\psi}\left\{\frac{2(n_{10} + n_{01})}{(2 + e^{\psi})^2} + \frac{2(n_{11} + n_{02})}{(1 + 2e^{\psi})^2}\right\}^{-1}. \qquad (3.15)$$

Numerical methods are required to obtain log odds ratio estimates for $c : 1$ matching where $c > 2$ and also in more complex cases involving multiple cases and controls in each matched set.

The precision to be achieved by $c : 1$ matching is most easily assessed by finding the Fisher information, that is, the reciprocal of the asymptotic variance of $\hat{\psi}$, evaluated at $\psi = 0$, the null hypothesis of no exposure difference between cases and controls. For this the possible data configurations are divided into sets depending on the total number of exposures, for which $X = 1$, and a conditional log likelihood for ψ is found within each set. These are differentiated twice with respect to ψ and then expectations taken over the sets. The resulting Fisher information at $\psi = 0$ is

$$\frac{c}{c+1} \sum_u L_1(\chi_u) L_0(\chi_u). \qquad (3.16)$$

It follows that the information from $c : 1$ matching relative to that from $1 : 1$ matching is $2c/(c + 1)$.

This result is equivalent to that for an unmatched study discussed in Section 2.5. Thus, in a matched case-control study with multiple controls matched to each case there is little to be gained in terms of efficiency by using more than about five controls per case.

To interpret (3.16) further, suppose that we write $m_\chi = \Sigma \chi_u/n$ and $v_\chi = \Sigma(\chi_u - m_\chi)^2/n$. Then, on taking two terms of the Taylor expansion

Table 3.5 *Estimated log odds ratio ($\hat{\psi}$), standard error (SE) and two-sided 95% confidence interval (CI) for colorectal cancer for fibre intakes of $\geq$ 15 grams/day as compared with < 15 grams/day, using $c = 1-4$ matched controls per case.*

| | Number of matched controls c | | | |
	1	2	3	4
$\hat{\psi}$	-0.170	-0.187	-0.284	-0.294
SE($\hat{\psi}$)	0.162	0.145	0.136	0.132
95% CI	$(-0.488, 0.148)$	$(-0.471, 0.097)$	$(-0.551, -0.017)$	$(-0.553, -0.035)$

about m_χ, (3.16) becomes

$$\frac{c}{c+1} n L_0(m_\chi) L_1(m_\chi) \left\{ 1 - \frac{v_\chi}{2} (6 L_0(m_\chi) L_1(m_\chi) - 1) \right\}. \qquad (3.17)$$

The corresponding information for the comparison of two independent binomial samples of sizes n and cn is

$$\frac{c}{c+1} n \pi (1 - \pi), \qquad (3.18)$$

where π is the common probability that $X = 1$.

Example 3.3: Number of matched controls In Example 3.1 we considered a matched-pair analysis of the data from a matched case-control study of fibre intake and colorectal cancer risk, ignoring the other three controls available for each case. We will extend this illustration by estimating the log odds ratio using two, three and four matched controls per case. The results are summarized in Table 3.5.

In this example it is only when we use three or four matched controls per case that a borderline significant protective effect of higher fibre intake on colorectal cancer risk is detected at the 0.05 significance level; as anticipated, the confidence intervals narrow gradually as c increases.

3.9 Logistic dependence

In Section 3.3 we introduced the *sampling model* $\Pr(X_{u;y} = 1) = L_1(\psi y + \chi_u)$ for a matched case-control study, where $X_{u;y}$ denotes the value of the explanatory variable ($X = 0, 1$) for an individual in the population generating the uth matched set; the outcome for the individual is given by $Y = y$ ($y = 0, 1$). This is the model for the data as they arise under matched case-control sampling. There is a corresponding representation for the *inverse model*. In the inverse model the labelling with respect to the fixed and random variables is reversed, and we let $Y_{u;x}$ denote the value of the outcome variable ($Y = 0, 1$) for an individual in the population generating the uth matched set with explanatory variable $X = x$ ($x = 0, 1$). For the inverse model we write

$$\Pr(Y_{u;x} = 1) = L_1(\psi x + \lambda_u). \tag{3.19}$$

As discussed in Section 2.4, χ_u and λ_u are quite different:

$$\chi_u = \pi_{u;10}/\pi_{u;00}, \quad \lambda_u = \pi_{u;01}/\pi_{u;00}.$$

Here we establish a connection with a more general account, needed to deal with situations of realistic generality. For this, a general formulation for the inverse model is required. This is in the form of a model for the binary variable Y_u, which defines whether an individual in the subpopulation generating matching set u is a case ($Y_u = 1$) or a control ($Y_u = 0$), given the explanatory variables of interest. In the general formulation of the inverse model, Y_u is assumed to have linear logistic regression on the vector X of explanatory variables, that is,

$$\Pr(Y_u = 1 | X = x) = L_1(\alpha_u + \beta^T x). \tag{3.20}$$

It will be shown in the next chapter that we can estimate the regression coefficients β under the sampling model for a matched case-control study.

Example 3.4: Indirect use of case-control methods   We illustrate the possibility of an extended use of the notion of a matched case-control study by outlining an investigation not originally conceived as a case-control study but in fact best analysed in that way. Jackson (2009) investigated whether employers discriminate between candidates for professional and managerial jobs on the basis of social class background characteristics. In this study speculative letters of application were sent to 2560 employers; each employer received letters from two hypothetical candidates with differing social class characteristics. Three such social class characteristics were used to indicate whether the applicant was in an 'elite' or in a 'non-elite' group: name, type of school attended and hobbies. For each of these three characteristics, Jackson identified an elite and a non-elite version. The letters differed in their style and wording to avoid detection but were substantively the same in terms of presentation and coherence. There were two outcomes of interest in this study, first, whether the applicant received a response and, second, whether the response was positive, in the form of an invitation to an interview or meeting.

In fact the proportion of positive responses was low. The exposure of interest was the social class of the applicant, as characterized above. Large differences between employers, for example in the nature of the employment likely to be available, the size of the organization involved and the attitude of the employer were to be expected but were not the direct object of study. This suggests the following parallel with a matched case-control study of a more conventional kind. A positive response is a case, a negative response a control. The employer is a basis of matching. Only employers yielding one case and one control, that is, one response and one non-response, provide information about the contrast of interest. Thus employers who answer either both or neither of the letters they receive provide no information about the effect of the social class of the applicant.

3.10 Introducing time

The incorporation of 'time' into the selection of controls was discussed in Section 2.10, where we considered dividing the period over which cases are ascertained into short time intervals. Here we will extend the discussion to matched analyses. As the time intervals within which cases and controls are sampled become very short or as, ideally, the control sampling occurs in continuous time, the resulting data is in the form of sets comprising one case and one or more controls from the risk set. The risk set at a given time is the set of individuals in the underlying population who are eligible to become a case at that time. We may therefore think of the case-control sets as being matched on 'time'. The matching might also extend to include other variables M.

For an individual at risk of becoming a case at time t we now define

$$\pi_{t;xy} = \Pr(X_t = x, Y_t = y), \tag{3.21}$$

where Y_t and X_t refer to the outcome and exposure status respectively of members of the population at risk at time t. For a case occurring instantaneously at time t, $\pi_{t;x1}$ is proportional to the probability density function at event time t for individuals with exposure x at t, which we denote by $f_x(t)$. For individuals at risk at t but not becoming cases at that time the probability $\pi_{t;x0}$ is proportional to the survivor function at time t or, equivalently, one minus the cumulative distribution function at t for individuals with exposure x at t. We denote this survivor function by $S_x(t)$. Note that

$$f_x(t) = h_x(t)S_x(t),$$

where $h_x(t)$ is the instantaneous hazard rate at time t for an individual with exposure x. The 'instantaneous' odds ratio at time t is therefore

$$\frac{\pi_{t;11}\pi_{t;00}}{\pi_{t;10}\pi_{t;01}} = \frac{h_1(t)}{h_0(t)}, \tag{3.22}$$

that is, the instantaneous odds ratio at time t is equal to a *hazard* or *rate ratio*.

Under the assumption that the hazard ratio in (3.22) does not change over time it follows from the above, and from the results in Section 3.3, that a conditional analysis is appropriate for a case-control study with controls selected using incidence density sampling and that the resulting estimated odds ratio is in fact an estimate of the hazard ratio. Many case-control studies use matching on age. If the matching is on date of birth, say to the nearest month, then this is a form of time matching with 'age' as the time scale.

A further advantage of the explicit introduction of time into the analysis is that it allows the exploration of time trends in exposure effects.

When this approach is used within an assembled cohort it is often referred to as a nested case-control study; nested case-control studies are discussed in detail in Chapter 7.

Example 3.1 continued The study first introduced in Example 3.1 had an element of time matching. Matched controls were required to have at least as much follow-up time as their case; that is, from the time of recruitment to the underlying cohort the controls were required to have been observed for at least as long as the case, without having been lost to follow-up or dying from another cause. The time scale in this situation was 'the time since recruitment'. A number of additional matching variables were also used.

Notes

Section 3.1. In discussions of the design of observational studies it may often be helpful to consider broadly parallel aspects of experimental design (Cochran, 1965). The matched-pair experimental design is a special case of a randomized block design, using one of the most basic methods of error control, that of comparing like with like, although the parallel may be far from complete.

Section 3.2. The choice of controls in a matched case-control study, including the choice of matching variables, the number of controls and overmatching, was discussed in detail by Wacholder *et al.* (1992b). Miettinen (1970) also commented on the choice of a matched or unmatched study and on some conditions for, and the implications of, using a matched study. Raynor and Kupper (1981) discussed some issues relating to the use of matching variables that are continuous, which requires matching to be done within categories of the continuous variable. Stuart (2010) reviewed matching techniques in a general setting.

Section 3.3. An analysis of matched binary pairs based only on discordant pairs was suggested for testing the null hypothesis of the absence of inter-treatment differences by McNemar (1947). For a theoretical derivation, including that of confidence intervals, see Cox (1958a, b), work summarized by Cox and Snell (1989). The use of exact inference for case-control studies is discussed in Hirji *et al.* (1988).

Section 3.5.2. Lynn and McCullogh (1992) considered how adequate the adjustment for covariates must be, in explaining the matching in a matched-pair study, before a regression adjustment analysis becomes more effective than a conditional analysis. Thomas and Greenland (1983) discussed statistical and practical issues relating to the choice of either a matched or an unmatched case-control design and to the choice of analysis in a matched study. Similar discussions are given also in Thompson *et al.* (1982) and in Kupper *et al.* (1981).

Section 3.7. Neyman and Scott (1948) described in general terms problems of the broad kind that arise under an unconditional analysis of matched case-control data, in which an attempt is made to estimate all the parameters χ_u. The expected bias in the log odds ratio estimate from this analysis, in matched case-control studies with multiple cases and controls in each matched set, was investigated by Pike *et al.* (1980). The bias reduces as the size of the matched sets increases, and it depends on the true odds ratio and on the proportion of exposed controls.

Section 3.8. The extension of conditional analysis to case-control studies with multiple controls per matched set was outlined by Miettinen (1969). Ury (1975) derived the statistical efficiency of $c : 1$ matched case-control studies relative to a matched-pair study, taking into account both binary explanatory variables and a continuous explanatory variable. The efficiencies are the same in both situations and are the same as those shown in Section 4.6. Factors influencing the optimal control-to-case ratio in a matched case-control study were investigated by Hennessy *et al.* (1999) for a binary explanatory variable and were found to depend on the ratio of Pearson's χ^2 statistic from the 2×2 table and the number of matched sets and on the prevalence of exposure among the controls. Dupont (1988) outlined power calculations for matched case-control studies.

Section 3.10. The estimation of hazard ratios in case-control studies using incidence density sampling was described by Prentice and Breslow (1978).

4

A general formulation

- Logistic regression can be used to estimate odds ratios using data from a case-control sample as though the data had arisen prospectively. This allows regression adjustment for background and confounding variables and makes possible the estimation of odds ratios for continuous exposures using case-control data.
- The logistic regression of case-control data gives the correct estimates of log odds ratios, and their standard errors are as given by the inverse of the information matrix.
- The logistic regression model is in a special class of regression models for estimating exposure-outcome associations that may be used to analyse case-control study data as though they had arisen prospectively. Another regression model of this type is the proportional odds model. For other models, including the additive risk model, case-control data alone cannot provide estimates of the appropriate parameters.
- Absolute risks cannot be estimated from case-control data without additional information on the proportions of cases and controls in the underlying population.

4.1 Preliminaries

The previous chapters have introduced the key features of case-control studies but their content has been restricted largely to the study of single binary exposure variables. We now give a more general development. The broad features used for interpretation are as before:

- a study population of interest, from which the case-control sample is taken;
- a sampling model constituting the model under which the case-control data arise and which includes a representation of the data collection process;
- an inverse model representing the population dependence of the response on the explanatory variables; this model is the target for interpretation.

Understanding the link between the sampling model and the inverse model is central to the estimation of exposure-outcome associations from case-control data. In Chapter 2 we showed that the log odds ratio of the outcome given a simple exposure in the study population is the same under the sampling model and under the inverse model. In the present chapter we extend this to outline the link between the sampling model and the inverse model under a general logistic regression model.

In accord with our previous notation, X denotes the main exposures of interest and Y denotes the outcome defining case or control status. The focus is on estimating the probability of the outcome $Y = 1$ (that is, of being a case) for an individual with exposure $X = x_1$ in comparison with that for an individual with exposure $X = x_2$, say, or more generally for a fixed change in X if it is continuous. We also define a set of variables W that are to be regarded as held fixed while the dependence of Y on X is assessed.

It was shown in Chapter 2 that, for a binary exposure X, the retrospective odds ratio,

$$\frac{\Pr(X = 1 | Y = 1, W = w) \Pr(X = 0 | Y = 0, W = w)}{\Pr(X = 0 | Y = 1, W = w) \Pr(X = 1 | Y = 0, W = w)}, \tag{4.1}$$

is equivalent to the target odds ratio,

$$\frac{\Pr(Y = 1 | X = 1, W = w) \Pr(Y = 0 | X = 0, W = w)}{\Pr(Y = 0 | X = 1, W = w) \Pr(Y = 1 | X = 0, W = w)}. \tag{4.2}$$

For this we made the crucial assumption that the distribution of X given (Y, W) is the same in the case-control sample as it is in the study population of interest.

Hence, for binary X the target exposure-outcome association, after conditioning on W, can be estimated from case-control data using the logistic regression of X on Y and W. Extensions of the above method to exposures X with several or many levels are possible, but they become increasingly clumsy and moreover are not applicable for continuous exposures X, which are often of interest.

4.2 Logistic regression for case-control studies

4.2.1 Initial formulation

As specified above we consider a target population in which there is an outcome or response Y, an exposure X and a vector W of variables that we wish to adjust for, all represented by random variables. It is assumed that the focus of interest is the conditional dependence of Y on X given W. Unmatched studies may use a component of W to perform stratified sampling and, in matched designs, a component M of W is used to define the matching. In this section we focus on unmatched studies in which the sampling of individuals depends only on the case-control status Y.

The ultimate object of study is, as already noted, the conditional distribution of Y given (X, W) in the study population of interest. This is, in our terminology, the *inverse model*, denoted by $\mathcal{I}$, and we provisionally assume that for $y = 0, 1$,

$$\Pr^{\mathcal{I}}(Y = y | X = x, W = w) = L_y(\alpha + \beta^T x + \gamma^T w), \qquad (4.3)$$

where the standardized logistic functions are given by

$$L_1(x) = \frac{e^x}{1 + e^x}, \quad L_0(x) = \frac{1}{1 + e^x}. \qquad (4.4)$$

Terms representing interactions between X and W could also be included in (4.3).

While the target distribution is as in (4.3), in a case-control study the distribution of the observations is determined by the sampling procedure for choosing cases and controls and by

$$\Pr^{\mathcal{D}}(X = x, W = w | Y = y), \qquad (4.5)$$

where the superscript $\mathcal{D}$ indicates conditioning on an individual being in the case-control sample, which we refer to henceforth as the sampling model; for continuous (X, W) (4.5) is a probability density function.

Our object is to relate the sampling model, essentially determining the likelihood of the data, to the inverse model representing the relations of interest in the population. There are a number of broadly equivalent ways of doing this. One is to specify at the outset the choice of fixed numbers of n_1 cases and of $n_0 = cn_1$ controls independently and randomly from the conditional distributions in the population, given respectively by $Y = 1, 0$. A slightly different route is related to the Poisson distribution-based formulation of Chapter 2. In this method we assume that cases and controls are sampled in independent Poisson processes with rates (ρ_1, ρ_0) and with a stopping rule conditionally independent of $V = (X, W)$. The contributions to the full likelihood from different individuals are then mutually independent. The full likelihood is a product of factors, that from individual i being

$$\mathrm{lik}_i = \Pr^{\mathcal{D}}(Y = y_i) \Pr^{\mathcal{D}}(V = v_i | Y = y_i). \qquad (4.6)$$

Using the assumption that the distribution of V given Y is the same in the case-control sample as it is in the underlying population, this can be rearranged in the form

$$\text{lik}_i = \frac{\Pr^{\mathcal{D}}(Y = y_i)\,\Pr^{\mathcal{I}}(Y = y_i | V = v_i)\,\theta_{v_i}^{\mathcal{I}}}{\Pr^{\mathcal{I}}(Y = y_i)} \tag{4.7}$$

where $\theta_v^{\mathcal{I}} = \Pr^{\mathcal{I}}(V = v)$, the distribution of V in the underlying population. When V is formed from binary indicators representing binary or categorical variables, the denominator can be written as

$$\sum_v \Pr^{\mathcal{I}}(Y = y_i | V = v)\theta_v^{\mathcal{I}}. \tag{4.8}$$

For continuous V the sum is replaced by an integral. Substitution of (4.3) into (4.7), and use of $\Pr^{\mathcal{D}}(Y = y) = \rho_y/(\rho_1 + \rho_0)$ gives in terms of (X, W)

$$\text{lik}_i = \frac{\rho_{y_i}}{\rho_1 + \rho_0} \frac{L_{y_i}(\alpha + \beta^T x_i + \gamma^T w_i)\theta_{v_i}^{\mathcal{I}}}{\sum_v L_{y_i}(\alpha + \beta^T x + \gamma^T w)\theta_v^{\mathcal{I}}}. \tag{4.9}$$

The second factor in (4.9) is of logistic form. Note that the latter is an extended version of the simple logistic models considered so far; it involves a sum of more than two terms in the denominator. Provided that the parameters (ρ_0, ρ_1) are variation independent of the regression parameters (β, γ) and that the corresponding sampling errors are asymptotically independent, we can regard the estimates of (β, γ) as holding conditionally on the estimate of ρ_1/ρ_0, namely n_1/n_0, which is targeted, at least approximately, to be $1/c$. This gives the individual likelihoods

$$\text{lik}_i \propto \frac{L_{y_i}(\alpha + \beta^T x_i + \gamma^T w_i)\theta_{v_i}^{\mathcal{I}}}{\sum_v L_{y_i}(\alpha + \beta^T x + \gamma^T w)\theta_v^{\mathcal{I}}}. \tag{4.10}$$

The full likelihood is the product over all individuals in the case-control sample. In simple situations the log odds ratio parameters β could be estimated by fitting a multinomial logistic model. However, unless X and W both contain only binary or categorical variables and there are not too many such variables, a different formulation is needed.

For this, we first note that the individual likelihood contribution in (4.6) or (4.7) can be expressed alternatively as

$$\text{lik}_i = \theta_{v_i}^{\mathcal{I}}\,\Pr^{\mathcal{I}}(Y = y_i | V = v_i)\,\eta_{y_i}, \tag{4.11}$$

where

$$\eta_y = \frac{\Pr^{\mathcal{D}}(Y = y)}{\Pr^{\mathcal{I}}(Y = y)} \tag{4.12}$$

expresses the distortion in the distribution of Y induced by the case-control sampling.

We now introduce $\theta_v^D = \Pr^D(V = v)$, which is the distribution of V in the case-control sample and can be written as

$$\theta_v^D = \sum_y \Pr^D(V = v|Y = y)\Pr^D(Y = y)$$

$$= \sum_y \Pr^I(V = v|Y = y)\Pr^D(Y = y)$$

$$= \theta_v^I \sum_y \Pr^I(Y = y|V = v)\,\eta_y, \tag{4.13}$$

using the assumption that the distribution of V given Y is the same in the case-control sample as in the underlying population. The term θ_v^I in (4.11) is now replaced by $\theta_v^I = \theta_v^D\{\sum_y \Pr^I(Y = y|V = v)\eta_y\}^{-1}$ from (4.13), giving

$$\text{lik}_i = \theta_{v_i}^D \frac{\Pr^I(Y = y_i|V = v_i)\,\eta_{y_i}}{\sum_y \Pr^I(Y = y|V = v_i)\,\eta_y}. \tag{4.14}$$

Finally we use the result that

$$\frac{L_y(\xi)}{L_0(\xi) + L_1(\xi)\eta_1/\eta_0} = L_y(\xi^*),$$

where $\xi^* = \xi + \log(\eta_1/\eta_0)$, together with the assumed logistic form of $\Pr^I(Y = y|V = v_i)$. This shows that, including the factor involving the distribution of V in the data, the log likelihood is given by

$$\text{lik} = \prod_i L_{y_i}(\alpha^* + \beta^T x_i + \gamma^T w_i) \prod_i \theta_{v_i}^D, \tag{4.15}$$

where

$$\alpha^* = \alpha + \log(\eta_1/\eta_0)$$
$$= \alpha + \log\left\{\Pr^I(Y = 0)/\Pr^I(Y = 1)\right\} + \log(\rho_1/\rho_0). \tag{4.16}$$

We now assume that the family of unknown distributions for V in the case-control sample is sufficiently rich that the corresponding observed marginal distribution on its own gives no useful information about (β, γ). With the other parameters unknown, all the information about (β, γ) is contained in the logistic regression component of the likelihood (4.15). As in the earlier discussion we can regard the estimates of ρ_y, which feature indirectly in α^*, as holding conditionally on the estimate of ρ_1/ρ_0, which is n_1/n_0. Because the contributions of different observations are independent, the standard large-sample properties of maximum likelihood theory apply; in particular the inverse information matrix estimates the precision matrix of the estimates. The conclusions of this discussion are that data arising from a case-control study can be analysed using the likelihood

$$\prod_i \Pr^{\mathcal{D}}(Y = y_i | X = x_i, W = w_i) = \prod_i L_{y_i}(\alpha^* + \beta^T x_i + \gamma^T w_i) \quad (4.17)$$

and that the estimates for the log odds ratio parameters β will be the same as would have been found had the data arisen prospectively.

A superficially more direct approach to the analysis, as noted above, is to start from samples of sizes n_1 and n_0 drawn independently and randomly without replacement from the subpopulations of cases and controls. Here, n_1 and n_0 are preassigned and fixed; in the Poisson-process approach they are not fixed. Since n_1 and n_0 are now fixed, the likelihood contributions in (4.17) cannot be treated as totally independent because different contributions to the score statistic are correlated and that correlation must be accounted for in assessing precision. In this case, therefore, (4.17) is a pseudo-likelihood and the variance-covariance matrix for the parameter estimates is given by the sandwich formula. Only the variance of the uninteresting parameter α^* is affected; the replacement of ρ_1/ρ_0 by n_1/n_0 reduces the apparently relevant variance by $1/n_0 + 1/n_1$. Further details on pseudo-likelihoods are given in the appendix.

Another way to arrive at the result (4.17) is to introduce constraints on the sizes of the samples of cases and controls, conditional on their being in the case-control sample. The full likelihood in (4.11) can be maximized subject to the constraints by the introduction of Lagrange multipliers, giving (4.17). This approach is discussed further in Chapter 6, where we consider two-stage designs.

4.2.2 Discussion

The above results show that, under a logistic model for Y, given (X, W) the same estimates of the log odds ratio parameter β are obtained whether one treats the case-control data as arising under the inverse model or under the sampling model. This is the central result enabling us to estimate odds ratios from a case-control study. As described thus far it depends on the logistic form for the inverse model in (4.3). Non-logistic forms for the inverse model are discussed in Section 4.5. Note that it is necessary also that the model specifies how cases and controls are chosen for study.

Throughout, the focus of interest is the dependence of Y on X given W. If X is a vector then the relations between the components of X in the population and in the case-control study will be different, but this is not of concern in the present context. If X_1 and X_2, for example, are independent in the population but both affect Y then, conditionally on Y, the components X_1 and X_2 are dependent, a dependence that is an artefact of the data-collection process.

If there is even a limited amount of external information about the marginal distribution of (X, W), the position is formally different from that described above. For example, if it were assumed that in the population X is multivariate normally distributed, then clear non-normality in the corresponding sample frequencies would be formal evidence of non-zero β. That is, the biased method of sampling based on case-control status will have changed the shape of the empirical distribution of X. One possibility is thus that independent information about the population probabilities $\theta^{\mathcal{I}}_{xw}$ is available, for example from a random sample of the population. Comparison with this might generate indirectly information about β. We return to this in Section 4.8.

4.2.3 A Bayesian approach

An alternative approach in terms of a Bayesian formulation requires the specification of a prior probability distribution that summarizes meaningfully information about the parameters other than that from the data themselves. For models with modest numbers of parameters the Bayesian approach with a suitable flat prior will give essentially the same numerical answers as the maximum likelihood approach. These ideas are reviewed briefly in appendix section A.8.

4.3 Extensions

So far we have supposed that cases and controls are sampled by some random process from the underlying populations with $Y = 1$ and $Y = 0$ respectively. In this section we extend the main results to two special sampling situations:

- the sampling of controls, or sometimes both cases and controls, within strata;
- the individual matching of controls to cases.

We begin by focusing on the first situation. We will assume that n_{1s} cases and n_{0s} controls are sampled in stratum s, for $s = 1, 2, \ldots$, notionally in independent Poisson processes; the strata are assumed to be based on one or more elements of the set of conditioning variables W.

We now assume that the inverse model is of the form

$$\Pr^{\mathcal{I}}(Y = y | X = x, S = s, W^* = w^*) = L_y(\alpha_s + \beta^T x + \gamma^T w^*), \quad (4.18)$$

where w^* specifies those parts of w, if any, that are not captured as inter-stratum effects α_s.

In this situation the full case-control likelihood is a product of terms, that from individual i being, with $V = (X, W^*)$,

$$\text{lik}_i = \Pr^{\mathcal{D}}(Y = y_i | W = w_i) \Pr^{\mathcal{D}}(V = v_i | Y = y_i). \quad (4.19)$$

The arguments in Section 4.2 can be extended directly to apply within strata, giving the result that case-control data obtained by stratified sampling can be analysed as though arising from a prospective sample maximizing the likelihood

$$\prod_i L_{y_i}(\alpha^*_{s_i} + \beta^T x_i + \gamma^T w^*), \quad (4.20)$$

where

$$\alpha^*_s = \alpha_s + \log\{\eta_1(w)/\eta_0(w)\}; \quad (4.21)$$

here $\eta_y(w) = \Pr^{\mathcal{D}}(Y = y | W = w)/\Pr^{\mathcal{I}}(Y = y | W = w)$.

The large-sample results of Section 4.2 also apply, with minor changes. Under the Poisson-based sampling scheme the variance-covariance matrix for $(\hat{\alpha}^*_1, \ldots, \hat{\alpha}^*_S, \hat{\beta})$ is the inverse information matrix I^{-1}. Under a scheme where the numbers of cases and controls sampled within strata are fixed, dependences are induced and the variance of $\hat{\alpha}^*_s$ obtained from the inverse information matrix can be corrected by reducing it by an amount $1/n_{1s} + 1/n_{0s}$, though these parameters are not of interest and the variance of $\hat{\beta}$ is unaffected.

Instead of estimating a separate intercept parameter α^*_s, stratification of the sampling may, in some circumstances, be accounted for by parame-terizing the effects of the variables used to define the sampling strata. In the above development the association between the variables W and the probability of outcome $Y = 1$ in the population, given by γ in (4.3), can be estimated only for those features W that are independent of the case-control sampling. For features W that are used to sample cases and controls the estimates of the associated parameters, γ^* say, do not have a useful

Table 4.1 *Possible outcomes in a matched-pair design and their conditional probabilities.*

Case	Control	Conditional probability	Value
0	0	$$\dfrac{L_1(\alpha_m^*)L_0(\alpha_m^*)}{L_1(\alpha_m^*)L_0(\alpha_m^*)}$$	1
1	0	$$\dfrac{L_1(\alpha_m^* + \beta)L_0(\alpha_m^*)}{\{L_1(\alpha_m^* + \beta)L_0(\alpha_m^*) + L_1(\alpha_m^*)L_0(\alpha_m^* + \beta)\}}$$	$\dfrac{e^\beta}{1 + e^\beta}$
0	1	$$\dfrac{L_1(\alpha_m^*)L_0(\alpha_m^* + \beta)}{\{L_1(\alpha_m^* + \beta)L_0(\alpha_m^*) + L_1(\alpha_m^*)L_0(\alpha_m^* + \beta)\}}$$	$\dfrac{1}{1 + e^\beta}$
1	1	$$\dfrac{L_1(\alpha_m^* + \beta)L_0(\alpha_m^* + \beta)}{L_1(\alpha_m^* + \beta)L_0(\alpha_m^* + \beta)}$$	1

interpretation. This approach is discussed further at the end of this section and is also expanded in Chapter 6.

Now consider the modification of the argument when controls are chosen to match each case individually with respect to variables M, whose components are components of W. For each case we select a fixed number c of controls from the study population. The case-control likelihood contribution in (4.17) is now best written in the modified form

$$\mathrm{Pr}^D(Y = 1 \mid X = x, M = m, W^* = w^*) = L_1(\alpha_m^* + \beta^T x + \gamma^T w^*),$$

$$(4.22)$$

where $\alpha_m^* = \alpha_m + \log\{\eta_1(m)/\eta_0(m)\}$ and $\eta_y(m) = \mathrm{Pr}^D(Y = y \mid M = m)/\mathrm{Pr}^{\mathcal{I}}(Y = y \mid M = m)$. Here α_m represents the effect of an individual's having matching feature m on the overall probability of the outcome and w^* specifies those parts of w, if any, that are not involved in the matching.

The contribution to the likelihood from a set of $c + 1$ individuals, comprising one case and c controls, is a product of factors all involving the same α_m^*. Where there are several components to the matching variables M, only a small number of individuals will usually make a contribution to estimation of α_m^*. This creates a challenge for the estimation of α_m^*. This issue was discussed in detail in Chapter 3. The most common solution is a *conditional* analysis. The objective is to obtain a statistical procedure, confidence limits or a significance test, whose statistical interpretation does not depend explicitly on the α_m^*.

In general, the conditional likelihood for a matched case-control study is

$$\text{lik}_{\text{CND}} = \prod_u \frac{\exp(\beta^T x_{u;1})}{\exp(\beta^T x_{u;1}) + \sum_{k=1}^c \exp(\beta^T x_{u;0k})}, \qquad (4.23)$$

where $x_{u;1}$ is the exposure vector for the case in the uth matched set and $x_{u;0k}$ is the exposure vector for the kth control in the uth matched set. This approach is referred to as *conditional logistic regression*.

An alternative for both stratified sampling and individually matched studies is for the effects of the sampling strata or of the matching variables on the outcome to be parameterized, for example as a linear logistic regression term. This is possible only if the features defining the strata or matched pairs can be summarized in a set of adjustment variables. For both types of study this would result in an unconditional logistic regression analysis with adjustment for the variables representing the strata or the matching. This requires assumptions about the functional form of the relationship. The simplest logistic models underpinning the two different analyses are not in general consistent with one another, however. This means in particular that comparison of the conclusions from the two analyses is not entirely straightforward. This is related to the *non-collapsibility* of odds ratios, which we discuss in the next section.

Example 4.1: Adjustment for multiple covariates   Here we give an illustration of the use of the general formulation to adjust for multiple covariates, in the setting of a matched case-control study of the association between fibre intake and colorectal cancer. This study, described in Example 3.1, is nested within the EPIC-Norfolk study. The general formulation introduced in this chapter allows additional adjustment in the matched analysis for other covariates that may potentially confound the association of interest. These are the components denoted by W^* in Section 4.3. The additional covariates considered were height, weight, smoking status (never-smoker, former smoker, current smoker), education level (four categories), social class (six categories), level of physical activity (four categories), average daily energy intake derived from fat and non-fat sources and average daily alcohol intake. The adjustment variables listed above account for 18 parameters in the logistic model. Without the use of the general formulation, simultaneous adjustment for these variables would not be feasible. All analyses assume the linearity of regression on a logistic scale.

The exposure of interest is average daily fibre intake, a continuous variable measured in units of 6 grams per day. Performing a conditional logistic regression analysis with adjustment for the covariates listed above gives an odds ratio 0.774 with standard error 0.084 and 95% confidence interval (0.626, 0.958).

4.4 More on non-collapsibility

The idea of non-collapsibility was introduced in Section 2.9. Recall that a parameter concerning the dependence of Y on X is said to be collapsible over W if (i) in the conditional dependence of Y on X given $W = w$ the parameter is the same at all values of w and (ii) in the marginal dependence of Y on X the parameter takes that same value. Odds ratios are, however, not collapsible.

We begin the discussion of the consequences with a preliminary result about the effect of averaging in a logistic model. To study this in its simplest form we consider $E\{L_1(Z)\}$, where Z is a random variable of mean μ and variance σ^2. For specific distributions for Z this could be calculated numerically but to obtain a general result we write $Z = \mu + \epsilon$, assume that ϵ is small and expand to the quadratic term in ϵ; this leads to the approximation

$$E\{L_1(Z)\} = L_1(\mu) + \sigma^2 L_1''(\mu)/2. \tag{4.24}$$

This can be written in different forms that are all equivalent to the same order of approximation and one such is obtained by incorporating the second term into the first to give

$$E\{L_1(Z)\} = L_1\left\{\mu + \sigma^2 L_1''(\mu)/(2L_1'(\mu))\right\}. \tag{4.25}$$

In fact

$$L_1''(\mu)/L_1'(\mu) = L_0(\mu) - L_1(\mu),$$

leading to the approximation

$$E\{L_1(Z)\} \approx L_1\{\mu + \sigma^2 D_{01}(\mu)/2\}, \tag{4.26}$$

where $D_{01}(\mu) = L_0(\mu) - L_1(\mu)$. The effect of this approximation is always to move the argument of the logistic function towards zero. There are many approximations equivalent to this one in their initial dependence on σ^2. One such approximation that has more stable behaviour for larger σ^2 is written as

$$L_1\left\{\frac{\mu}{(1 - D_{01}(\mu)\sigma^2/\mu)^{1/2}}\right\}. \tag{4.27}$$

Note that close to $\mu = 0$ the value of $D_{01}(\mu)/\mu$ is $-1/2$. This in turn is closely related to yet another approximation, obtained by replacing the logistic function by a normal integral, assuming the random variable ϵ to be normally distributed and then evaluating the expectation in closed form.

We now apply the approximation (4.26) to study the consequences for relations such as

$$\mathrm{Pr}^{\mathcal{I}}(Y = 1|X = x, W = w) = L_1(\alpha + \beta_{YX.W}x + \beta_{YW.X}w),$$

where $\beta_{YX.W}$ is the change in the log odds of Y per unit change in X with W fixed (as implied by the subscript point) and similarly for $\beta_{YW.X}$; we consider one-dimensional X and W for simplicity. Thus in the underlying population Y depends on both X and W and the parameters β define conditional dependences. Suppose now that we wish to marginalize over W, assuming for simplicity that W has a linear regression on X,

$$W = \mu_W + \beta_{WX}X + \epsilon_{W.X}, \tag{4.28}$$

where $\epsilon_{W.X}$ has zero mean, is independent of X and has variance σ_ϵ^2.

It follows that, marginalizing over W,

$$\mathrm{Pr}^{\mathcal{I}}(Y = 1|X = x)$$
$$= E\{L_1(\tilde{\alpha} + (\beta_{YX.W} + \beta_{YW.X}\beta_{WX})x + \beta_{YW.X}\epsilon_{W.X})\}. \tag{4.29}$$

The expectation can be found approximately from (4.26) provided that the contribution of $\epsilon_{W.X}$ is relatively small. We have that

$$\mathrm{Pr}^{\mathcal{I}}(Y = 1|X = x)$$
$$= L_1\left(\tilde{\alpha} + \beta_{YX}x + D_{01}(\tilde{\alpha} + \beta_{YX}x)\beta_{YW.X}^2\sigma_\epsilon^2/2\right). \tag{4.30}$$

Here

$$\beta_{YX} = \beta_{YX.W} + \beta_{YW.X}\beta_{WX}, \tag{4.31}$$

often called Cochran's formula in the context of linear least squares regression, where it applies exactly.

There are a number of implications of this result, which we now consider briefly.

The logistic regression of Y on X is quite different from that of Y on X given W unless Y has zero regression on W given X, $\beta_{YW.X} = 0$. This is true even if W has zero regression on X, $\beta_{WX} = 0$, unless σ_ϵ^2 is small. This is due to the effect of the correction term in (4.26) or (4.30). Marginalizing over W tends to shrink the regression coefficients by an amount (which may, however, be small) depending on the variability associated with W. A numerical example was given in Example 2.2, and some implications of these results were given in Section 2.9.

There are also implications for matched studies, which include the following:

- matched and unmatched analyses, either of the same data or of closely related sets of data, will not in general yield the same results, even when the matching variables are not confounders;
- related is that a conditional logistic analysis of a matched study will give different results from with an analysis which deals with the matching variables using an unconditional regression adjustment analysis;
- similarly, a stratified analysis allowing a different intercept for each stratum will give different results in comparison with an analysis in which the strata are represented by a linear logistic regression relation.
- Similarly, if stratification or matching are used, account of this should be taken in analysis of the associations of primary interest.

Example 4.2: Properties of odds ratios Here we compare odds ratio estimates obtained under a conditional matched analysis with the corresponding unconditional regression adjustment analysis.

The UK national case-control study of cervical cancer was established to investigate risk factors for cervical cancer in women in the UK. See Green *et al.* (2003) for details. Cases were selected from five UK cancer registries and divided into two types, squamous cell carcinoma cases and adenocarcinoma cases. For each squamous cell carcinoma case, one control was selected from the same region of residence of the case (using GP practice) and with the same age as the case within 12 months. Each adenocarcinoma case was matched to three controls using the same matching criteria. Information was obtained on a range of exposures, including reproductive factors, cervical screening, body weight, smoking, occupation and the use of hormones.

In the regression adjustment analysis the matching variables were accounted for by adjustment for age and region of residence. Odds ratios were estimated for each exposure, continuous exposures being divided into groups. In total, we estimated 45 odds ratios under both methods of analysis. All models were adjusted for known or suspected confounders that were not matching variables. Squamous cell carcinoma cases and adenocarcinoma cases were considered separately. In Figure 4.1 we show plots of the log odds ratio estimates under the regression adjustment analysis versus those from the matched analysis.

The odds ratios are not identical under the two approaches, though they are very similar. There is some suggestion that the regression adjustment estimates are marginally closer to the null compared with those from the matched analysis.

(a) Squamous cell carcinoma cases and 1 : 1 matched controls.

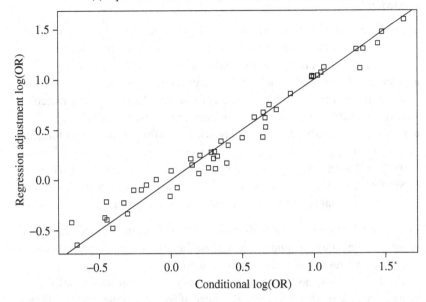

(b) Adenocarcinoma cases and 3 : 1 matched controls.

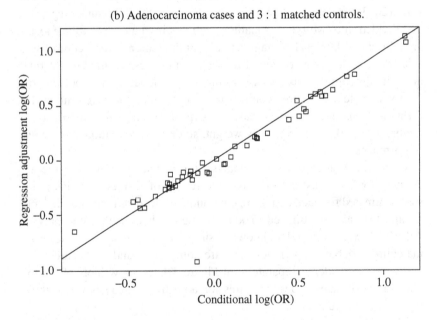

Figure 4.1 The UK national case-control study of cervical cancer. The estimated odds ratios from a matched analysis of case-control data versus the corresponding unmatched analysis with regression adjustment for the matching variables.

4.5 Non-logistic models

4.5.1 Multiplicative intercept models

The discussion so far has focused on a population for which there is an assumed linear logistic relation between case-control status and the explanatory variables, that is an inverse model of the logistic form given in (4.3). Under a logistic form for the inverse model $\Pr^{\mathcal{I}}(Y = y|X = x)$, the sampling model $\Pr^{\mathcal{D}}(Y = y|X = x)$, which conditions on the case-control sampling procedure, will also be logistic. Crucially, the odds ratio is the same under the two models and the odds of the outcome $Y = y$ ($y = 0, 1$) are the same up to a multiplicative constant:

$$\frac{\Pr^{\mathcal{I}}(Y = y|X = x)}{1 - \Pr^{\mathcal{I}}(Y = y|X = x)} = e^{\alpha} e^{\beta^T x}, \tag{4.32}$$

$$\frac{\Pr^{\mathcal{D}}(Y = y|X = x)}{1 - \Pr^{\mathcal{D}}(Y = y|X = x)} = e^{\alpha^*} e^{\beta^T x}. \tag{4.33}$$

In fact, the logistic model is one of a special class of so-called *multiplicative intercept* models, under which it can be shown that the log odds of the outcome $Y = y$ is the same up to a multiplicative constant. That is, for any multiplicative intercept model, $\Pr^{\mathcal{D}}(Y = y|X = x)$ and $\Pr^{\mathcal{I}}(Y = y|X = x)$ will be of the same form and the parameters representing the effects of X on Y will be the same under the two models. For a binary outcome, the general form for the models above under a multiplicative intercept inverse model is

$$\Pr^{\mathcal{I}}(Y = 1|X = x) = \frac{\phi r(x; \beta)}{1 + \phi r(x; \beta)}, \tag{4.34}$$

$$\Pr^{\mathcal{D}}(Y = 1|X = x) = \frac{\phi^* r(x; \beta)}{1 + \phi^* r(x; \beta)}, \tag{4.35}$$

where $r(\cdot)$ is some function of x. Under the logistic model we have $\phi = e^{\alpha}$ and $r(x; \beta) = e^{\beta^T x}$.

Other models in this class include the proportional odds model for ordered exposures, also known as the ordered logistic model, and the linear odds model, which is discussed in the example below. In these, in effect linear logistic regression is replaced by special forms of nonlinear logistic regression.

Example 4.3: Use of linear odds model Fearn *et al.* (2008) reported on a combined study of a number of case-control studies of residential radon exposure and lung cancer, covering several countries. The exposure measurement of interest, X, was average radon exposure over the previous 30 years. Individuals were classified into a large number of strata, denoted by S, defined by study, age, sex, region of residence and smoking history. The analysis was based on a *linear odds* model of the form

$$\Pr^{\mathcal{D}}(Y = 1 | X = x, S = s) = \frac{\alpha_s^*(1 + \beta x)}{1 + \alpha_s^*(1 + \beta x)}. \tag{4.36}$$

The linear odds model has been found to be more suitable for studies of radiation as a risk factor. The odds ratio under this model is $1 + \beta x$, assuming a comparison with zero exposure. A conditional analysis was used to eliminate the large number of nuisance parameters α_s^*.

4.5.2 Further discussion

We now examine more generally the consequences of a non-logistic relation for the inverse model. We write

$$\Pr(Y = y | X = x) = a_y \left(\alpha_{NL} + \beta_{NL}^T x \right), \tag{4.37}$$

where the subscript NL denotes parameters in a *non-logistic* model. For example, if the population relation is linear then $a_1(t) = t$ and the probability equation (4.37) is linear.

The logistic representation of the relation will not be linear. We write

$$\Pr(Y = 1 | X = x) = L_1 \left(\alpha_L + \beta_L^T x \right) \left\{ 1 + \epsilon g \left(\alpha_L + \beta_L^T x \right) L_0 \left(\alpha_L + \beta_L^T x \right) \right\}, \tag{4.38}$$

where the subscript L denotes parameters in a *logistic* model. Here the function $g(\cdot)$ characterizes the departure from linear logistic form and ϵ is a notional small quantity used to develop results locally near the null hypothesis $\beta = 0$.

It follows from (4.38) that, on neglecting terms in ϵ^2,

$$\log \frac{\Pr(Y = 1 | X = x)}{\Pr(Y = 0 | X = x)} = \alpha_L + \beta_L^T x + \epsilon g \left(\alpha_L + \beta_L^T x \right). \tag{4.39}$$

That is, if it is assumed that there is a linear dependence on some unknown scale then a departure from linearity on the logistic scale can be used to estimate the unknown scale in question.

Note that, without approximation,

$$\epsilon g \left(\alpha_L + \beta_L^T x \right) = \frac{\Pr(Y = 1 | X = x) - L_1(\alpha_L + \beta_L^T x)}{L_1 \left(\alpha_L + \beta_L^T x \right) L_0 \left(\alpha_L + \beta_L^T x \right)}. \tag{4.40}$$

The parameters (α_L, β_L) and $(\alpha_{NL}, \beta_{NL})$ in the relations (4.37) and (4.40) are numerically different but in any particular case can be chosen to produce approximate agreement over any specified range. For an outline general discussion, suppose that X is defined in such a way that $x = 0$ represents a standard reference condition, for example males of median age. We then argue locally near $\epsilon = 0$. For the best local agreement between (4.37) and the approximating logistic relation, we equate the values of and the tangents to the two models at the reference point to give

$$a_1(\alpha_{NL}) = L_1(\alpha_L), \quad a'_1(\alpha_{NL})\beta_{NL} = L'_1(\alpha_L)\beta_L. \qquad (4.41)$$

It is important that here the parameters α refer to the population model, not to the sampling model used to analyse the case-control data.

If, in particular, the initial model is linear then $a_1(t) = t$, so that

$$\alpha_{NL} = \frac{e^{\alpha_L}}{1 + e^{\alpha_L}}, \quad \beta_{NL} = \frac{e^{\alpha_L}}{(1 + e^{\alpha_L})^2}\beta_L. \qquad (4.42)$$

In a typical case-control context, in the population the probability that $Y = 1$ is small, so that approximately

$$\alpha_{NL} = e^{\alpha_L}, \quad \beta_{NL} = e^{\alpha_L}\beta_L. \qquad (4.43)$$

Thus, to convert the logistic regression coefficient into an estimate on the linear-in-probability scale, or indeed on any other scale, essentially requires evidence in this situation about the probability of a case at $x = 0$.

4.6 Statistical efficiency

A central question concerns the relative advantages of matched and un-matched studies. The choice between a matched-pair configuration and a comparison of two independent samples involves issues that for approximately normally distributed observations are different from those for the binary data arising in case-control studies. For the former, typically the target of the matched design is the reduction, possibly by a substantial factor, in the effective error variance. In the latter the issue is more subtle. The logistic parameter comparing cases and controls in the latter defines a log odds ratio conditionally on the features M defining the matching. In a corresponding unmatched design, in which controls are chosen randomly from the relevant population, the logistic parameter is the log odds ratio comparing probabilities marginalized over M. Because of possible confounding by M and because of the non-collapsibility issue the two parameters are not the same except when their common value is zero, although the difference may be quite small, in particular if there is no confounding by M.

The strong appeal in many contexts of a matched analysis studied by conditional logistic regression lies largely in its ability to deal simply with a range of features M without too detailed a model for the dependence.

We now turn to a comparison of the statistical efficiencies in matched and unmatched studies. It is convenient to return at this point to the logistic form for the *sampling model* for a binary exposure and outcome. The sampling model for the uth matched set is

$$\Pr(X_{u;y} = 1) = L_y(\beta y + \chi_u). \tag{4.44}$$

We consider a study in which n cases have been individually matched each to c controls, and there is a binary exposure X. Suppose first that the χ_u are all equal. In this situation the matching is not necessary to obtain an unconfounded estimate of the exposure-outcome association. This suggests that a conditional analysis may result in a loss of efficiency relative to an unconditional analysis, in particular because of the rejection of the concordant pairs. This is most easily investigated by direct computation of the asymptotic variances in terms of the underlying parameters of the sampling model in (4.44). Under an unmatched study the asymptotic variance of the estimate of β is

$$\frac{1}{cne^{\chi+\beta}}\{c(1 + e^{\chi+\beta})^2 + e^{\beta}(1 + e^{\chi})^2\}. \tag{4.45}$$

By contrast, in a conditional analysis the large-sample variance of an estimate based on the discordant sets can be written in terms of parameters as

$$n^{-1}\left\{\sum_{k=0}^{c}\binom{c}{k}e^{(k-1)\chi}L_1(\chi+\beta)L_1(\chi)^c\right.$$
$$\left.\times\left(\frac{(k+1)(c-k)e^{\chi+\beta}}{\{(k+1)e^{\beta}+(c-k)\}^2} + \frac{k(c-k+1)}{\{ke^{\beta}+(c-k+1)\}^2}\right)\right\}^{-1}, \tag{4.46}$$

using the result that the expected numbers of pairs with respectively the case exposed and unexposed, and k controls exposed, are given by

$$n\binom{c}{k}L_1(\chi+\beta)L_1(\chi)^k L_0(\chi)^{c-k}, \tag{4.47}$$

$$n\binom{c}{k}L_0(\chi+\beta)L_1(\chi)^k L_0(\chi)^{c-k}. \tag{4.48}$$

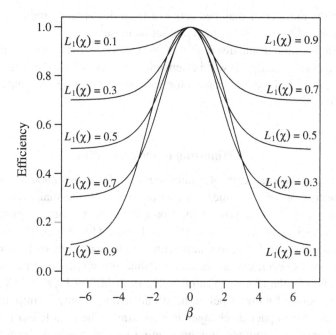

Figure 4.2 Asymptotic relative efficiency of the conditional
analysis relative to the unconditional analysis for a pair-matched
case-control study when $\chi_m = \chi$ for all pairs m; $L_1(\chi) =$
$e^\chi/(1 + e^\chi)$.

Under a pair-matched study ($c = 1$) the asymptotic relative efficiency
of the conditional analysis relative to the, in this case, more appropriate
unconditional analysis is the ratio of (4.46) and (4.45), giving

$$\frac{e^\beta(1 + e^\chi)^2 + (1 + e^{\chi+\beta})^2}{(1 + e^\chi)(1 + e^\beta)(1 + e^{\chi+\beta})}. \tag{4.49}$$

Locally near $\beta = 0$ this ratio is unity, as is the corresponding efficiency
in the more general case of $c : 1$ matching. That is, there is no loss of
asymptotic efficiency in using the conditional analysis when the matching
is in fact ineffective. This perhaps surprising result shows that, at least near
the null hypothesis that cases and controls have the same distribution of X,
the concordant pairs contain little or no useful information about β.

Figure 4.2 shows the asymptotic relative efficiency across a range of
values for the log odds ratio β under $1 : 1$ matching. The loss of efficiency in
the conditional analysis due to random matching is less severe as c increases

because the number of matched sets lost from discordancy reduces as the number of individuals in the matched set increases.

The implication of the discussion in this section is that in most matched-pair case-control studies it will be sensible to use primarily the conditional analysis, even if there are reasons for thinking the pairing may have been largely ineffective.

4.7 Estimating other quantities

We have seen that case-control studies provide information about the effects of exposures on an outcome, via an odds ratio. While many studies of associations between exposures and outcomes are focused primarily on *relative* associations, it is also sometimes of interest to estimate the *absolute* risk of an outcome for individuals with certain exposures or to compare associations between exposures and outcomes on an absolute probability or frequency scale, that is, to estimate the probabilities $Pr(Y = y|X = x)$. From the point of view of public health decisions, it may be important to understand the impact of changes in exposure at the population level on financial or other resources, and this requires an assessment of the numbers of individuals affected. Absolute risks are also important when offering advice to individuals on whether to take action to change their exposure, if this is possible, for example whether to give up smoking.

Two particular quantities that are sometimes considered are the *risk difference* and the *attributable risk*. For a binary exposure the risk difference is

$$\Gamma = Pr(Y = 1|X = 1) - Pr(Y = 1|X = 0). \qquad (4.50)$$

The attributable risk is the proportion of cases that can be attributed to a given level of exposure in the population or to an exposure above or below some threshold level for continuous exposures. For a binary exposure the risk of outcome $Y = 1$ attributable to an exposure $X = 1$ is

$$\Gamma_a = \{Pr(Y = 1) - Pr(Y = 1|X = 0)\}/Pr(Y = 1). \qquad (4.51)$$

Case-control studies do not, on their own, provide estimates of the probabilities $Pr(Y = y|X = x)$ and therefore the above quantities cannot be estimated without additional information. One possibility considered here is that in addition to the case-control data there is available some information on the total numbers of cases, N_1, and controls, N_0, in the underlying population, or perhaps the ratio of the number of cases and the number of controls. Such information may be readily available if the case-control study is nested within a cohort, or may be available from an external source.

We assume, as before, a logistic inverse model, namely $\Pr^{\mathcal{I}}(Y = 1 | X = x) = L_1(\alpha + \beta x)$. We have established that the log odds ratio parameters β from the inverse model can be estimated from a case-control study, but that we need to estimate α^* instead of α, where $\Pr^{\mathcal{D}}(Y = 1 | X = x) = L_1(\alpha^* + \beta x)$. To estimate an absolute risk requires the estimation of α. Earlier we found the relation

$$\alpha^* = \alpha + \log(\eta_1/\eta_0),$$

where $\eta_y = \Pr^{\mathcal{D}}(Y = y)/\Pr^{\mathcal{I}}(Y = y)$ is the ratio of the probability of $Y = y$ in the data and the probability of $Y = y$ in the population. As given in Section 4.2.1, the ratio $\Pr^{\mathcal{D}}(Y = 1)/\Pr^{\mathcal{D}}(Y = 0)$ can be estimated by n_1/n_0. Using additional information about the total numbers of cases and controls in the underlying population, $\Pr^{\mathcal{I}}(Y = 1)/\Pr^{\mathcal{I}}(Y = 0)$ can be estimated by N_1/N_0. It follows that α is estimated by

$$\hat{\alpha} = \hat{\alpha}^* - \log\{(n_1 N_0)/(n_0 N_1)\}. \tag{4.52}$$

When the population totals N_1 and N_0 are known, not estimated, the variance of the estimate of $\hat{\alpha}$ is equal to the variance of $\hat{\alpha}^*$; this was discussed in Section 4.2. On the other hand, if the proportions of cases and controls in the underlying population are estimated in a sample from that population then the additional variability in N_1 and N_0 must be taken into account in the variance by increasing it by a factor $1/N_1 + 1/N_0$, the variance of $\log N_1 - \log N_0$.

Whenever we can estimate absolute risks we can estimate also attributable risk. For a single binary exposure the attributable risk as given in (4.51) can be estimated using

$$\hat{\Gamma}_a = \frac{N_0(r_{00} n_1 - r_{01} n_0)}{r_{01} n_0 N_1 + r_{00} n_1 N_0}, \tag{4.53}$$

where r_{xy} denotes the number of individuals in the case-control sample with $(X = x, Y = y)$.

Under some further assumptions the attributable risk can be estimated using case-control data alone. To see this, we first rearrange (4.51) to give an alternative expression,

$$\Gamma_a = 1 - \Pr(X = 0 | Y = 1)/\Pr(X = 0). \tag{4.54}$$

If the outcome $Y = 1$ is rare in the underlying population then we have that, approximately, $\Pr(X = 0) = \Pr(X = 0 | Y = 0)$, giving

$$\Gamma_a^* = 1 - \Pr(X = 0 | Y = 1)/\Pr(X = 0 | Y = 0). \tag{4.55}$$

Assuming that $\Pr(X = x|Y = y)/\Pr^{\mathcal{D}}(X = x|Y = y)$, the attributable risk can then be estimated directly from a case-control study, and in the binary case the estimate is

$$\hat{\Gamma}_a^* = \frac{r_{11}r_{00} - r_{01}r_{10}}{r_{.1}r_{00}}. \qquad (4.56)$$

The variance of $\hat{\Gamma}_a^*$ can be estimated, for example by using their relation to the Poisson-distributed random variables set out in Section 2.3.2.

4.8 Use of the distribution of exposure

In the previous section we considered the possibility that case-control data may be supplemented with estimates of the proportions of cases and controls in the underlying population. Suppose that instead we have information about the distribution of the exposure in the underlying population. We consider briefly how this information can be used.

Information on the marginal distribution of exposure alone is not sufficient to allow estimation of absolute risks, since for this some information about the marginal distribution of the outcome is necessary. However, additional information about the marginal distribution of exposure alone can be used, formally at least, in improved estimation of the log odds ratio parameter β from case-control study data.

For this we consider for simplicity binary X and we ignore W. A typical contribution to the likelihood is proportional to

$$
\begin{aligned}
\Pr^{\mathcal{D}}(X = x_i|Y = y_i) &= \Pr^{\mathcal{I}}(X = x_i|Y = y_i) \\
&= \frac{\Pr^{\mathcal{I}}(Y = y_i|X = x_i)\theta_{x_i}^{\mathcal{I}}}{\Pr^{\mathcal{I}}(Y = y_i)} \\
&= \frac{L_{y_i}(\alpha + \beta x_i)\theta_{x_i}^{\mathcal{I}}}{\theta_0^{\mathcal{I}} L_{y_i}(\alpha) + \theta_1^{\mathcal{I}} L_{y_i}(\alpha + \beta)}.
\end{aligned}
$$

We now simplify the argument by supposing that cases are rare, that is, that e^{α} is small. Then for controls with $X = x_i$ the contribution to the likelihood is $\theta_{x_i}^{\mathcal{I}}$, whereas for those cases with $x_i = 1$ it is

$$\frac{e^{\beta}\theta_1^{\mathcal{I}}}{\theta_0^{\mathcal{I}} + e^{\beta}\theta_1^{\mathcal{I}}}. \qquad (4.57)$$

For cases with $x_i = 0$ the contribution is

$$\frac{\theta_0^{\mathcal{I}}}{\theta_0^{\mathcal{I}} + e^{\beta}\theta_1^{\mathcal{I}}}. \qquad (4.58)$$

If we regard the $\theta_x^{\mathcal{I}}$ as known, it follows that the log likelihood is

$$r_{11}\beta - r_{.1}\log(\theta_0^{\mathcal{I}} + \theta_1^{\mathcal{I}}e^{\beta}),\tag{4.59}$$

where, using our previous notation, r_{xy} denotes the number of individuals in the case-control sample with $(X = x, Y = y)$.

Maximization then gives the estimate, in fact a risk ratio,

$$e^{\tilde{\beta}} = \frac{r_{11}\theta_0^{\mathcal{I}}}{r_{01}\theta_1^{\mathcal{I}}}.\tag{4.60}$$

That is, the contribution of the control group is replaced by population proportions assumed to be known from an independent source.

Allowance for independent sampling errors in the population values can be made, in principle by adding a further term to the likelihood. If these external values had very high precision and could be used with security then the sampling of controls within the study would be unnecessary. In practice, depending on the effort involved, some within-study controls might well be desirable, in particular as a partial check on the relevance of the external data. There is a connection here with case-only studies, which are discussed in Chapter 5.

A complementary issue connected with the distribution of exposure is that it is not legitimate to study the interrelation between the distribution of different components of the exposure directly from their overall distribution in the case-control data. The reason is that the strong dependence of becoming a case on a particular component, say x^*, of the exposure will increase, often substantially, the general level of x^* in the cases. If cases are rare in the population then the distribution of exposure in the controls may be sufficiently close to that in the population to be used directly. In other situations separate conditional distributions in cases and controls may form the basis of a synthesis.

4.9 Beyond case-control studies: a more general view

The main theme of this book concerns retrospective sampling from a binary outcome, thus comparing cases with controls. We now outline briefly a more general discussion of retrospective sampling, taking an initially arbitrary pair of continuous random variables X and Y representing in the underlying population the explanatory variable and the outcome respectively. Denote their joint density in the population by $f_{XY}(x, y)$ with a corresponding notation for the conditional and marginal densities. That is, we write

$$f_{XY}(x, y) = f_{Y|X}(y; x) f_X(x) = f_{X|Y}(x; y) f_Y(y). \qquad (4.61)$$

On taking logs and differentiating, we have that

$$\frac{\partial^2 \log f_{XY}(x, y)}{\partial x \partial y} = \frac{\partial^2 \log f_{Y|X}(y; x)}{\partial y \partial x} = \frac{\partial^2 \log f_{X|Y}(x; y)}{\partial x \partial y}. \qquad (4.62)$$

The first of these three versions may be called the *local log odds function*, because for any function $g(x, y)$ the mixed second derivative is proportional to the limit as $h \to 0$ of

$$\{g(x+h, y+h) - g(x-h, y+h) - g(x+h, y-h) + g(x-h, y-h)\} \frac{1}{h^2}.$$

By a similar argument,

$$\frac{\partial^2 \log f_{Y|X}(y; x)}{\partial y \partial x} \qquad (4.63)$$

gives the local log odds for Y by specifying the rate of change with respect to x in the ratio of the densities at values $y + h$ and $y - h$, conditionally on $X = x$.

Thus (4.62) expresses the equality of log odds ratios in retrospective sampling, in prospective sampling and in the population. Thus in principle, if not in practice, if a nonparametric estimate of the second derivative could be found from a sufficiently rich form of retrospective sampling then an estimate of the corresponding prospective function would be achieved. We shall instead proceed parametrically.

Consider the special case where the local log odds function is constant, that is,

$$\frac{\partial^2 \log f_{X,Y}(xy)}{\partial x \partial y} = \gamma, \qquad (4.64)$$

leading to

$$f_{X,Y}(x, y) = a(x) b(y) e^{\gamma xy}, \qquad (4.65)$$

where $a(.)$, $b(.)$ are arbitrary non-negative functions subject to a normalizing condition. It follows that

$$f_{X|Y}(x; y) = \frac{a(x) e^{\gamma xy}}{a^*(\gamma y)}, \quad f_{Y|X}(y; x) = \frac{b(y) e^{\gamma xy}}{b^*(\gamma x)}.$$

Here, for example, the normalizing constant $a^*(\gamma y)$ is, except for the sign of its argument, the Laplace transform of $a(\cdot)$.

In a fully parametric formulation the functions $a(\cdot)$, $b(\cdot)$, determining respectively the marginal distributions of X and of Y, would typically be specified by variation-independent parameters. It is then clear that from a retrospective sample alone the most that can be estimated about the prospective dependence is the parameter γ. If, however, some information about the marginal distribution of Y, that is, information about $b(\cdot)$, is available from additional data then more can be estimated about the prospective dependence. This is consistent with the conclusion in the simple case-control study, where, for example, the population marginal proportion of cases cannot be estimated, only the odds ratio.

As an explicit special case, suppose that the population distribution of (X, Y) is bivariate normal with variances (σ_X^2, σ_Y^2) and correlation ρ. We assume for simplicity that the means are either known or of no interest. The joint density, without the normalizing constant, is written first as a product, namely

$$\exp\left\{-\frac{x^2}{2(1-\rho^2)\sigma_X^2}\right\} \exp\left\{\frac{\rho xy}{(1-\rho^2)\sigma_X\sigma_Y}\right\} \exp\left\{-\frac{y^2}{2(1-\rho^2)\sigma_Y^2}\right\}.$$

This is of the form (4.65) with

$$\gamma = \frac{\rho}{\sigma_X\sigma_Y(1-\rho^2)}, \quad a(x) = \exp\left\{-\frac{x^2}{2(1-\rho^2)\sigma_X^2}\right\} \tag{4.66}$$

and an analogous form for $b(\cdot)$. Then in a prospective study the primary parameter of interest would be the linear regression coefficient of Y on X, namely $\beta_{YX} = \rho\sigma_Y/\sigma_X$. In a retrospective study we would estimate for two or more values of Y aspects of the conditional density of X given Y, yielding thereby estimates of $\beta_{XY} = \rho\sigma_X/\sigma_Y$ and the conditional variance $\sigma_{X.Y}^2 = \sigma_X^2(1-\rho^2)$. If we denote the corresponding estimates by $\hat{\beta}_{XY}$ and $\hat{\sigma}_{X.Y}^2$ respectively, it follows that $\hat{\beta}_{XY}/\hat{\sigma}_{X.Y}^2$ estimates γ as defined by (4.66). By symmetry, the same parameter can be estimated by a prospective study, that is without estimating the marginal distribution of X. Unfortunately it seems unlikely that γ can fruitfully be interpreted in subject-matter terms, except perhaps for comparing two or more sets of data in which the marginal variances may reasonably assumed constant.

To proceed, it is necessary to estimate separately a further parameter. In any context in which retrospective sampling is relevant it is Y rather than X that is most accessible, and so we suppose that an independent estimate of the marginal variance σ_Y^2 is available. From $(\hat{\beta}_{XY}, \hat{\sigma}_{X.Y}^2, \hat{\sigma}_Y^2)$ all three parameters of the covariance matrix of (X, Y) can be estimated by equating these three statistics to their expectations and solving. For example, ρ is estimated by

$$\frac{\hat{\sigma}_Y \hat{\beta}_{XY}}{(\hat{\sigma}_{X.Y}^2 + \hat{\sigma}_Y^2 \hat{\beta}_{XY}^2)^{1/2}}.$$

Another special choice is, after scale changes, $a(x) = b(x) = ce^{-x}$ for $x > 0$, where c is a normalizing constant. A further choice is to take $a(\cdot)$ and $b(\cdot)$ a constant over finite intervals and zero otherwise.

Thus, in principle at least, studies of the interrelation between distinct ways of sampling a population are by no means confined to binary data and logistic models.

Notes

Section 4.1. This older approach was discussed in detail by Prentice (1976), who surveyed other previous work.

Section 4.2. The discussion here largely follows Prentice and Pyke (1979) and Farewell (1979). It built on a similar argument for discrete exposures given by Anderson (1972). An early discussion focusing on covariate adjustment was given by Breslow and Powers (1978). A similar development was also given by Cosslett (1981) from the point of view of choice-based sampling in econometric application and was not restricted to discussion of the logistic model. The Poisson sampling scheme is just one way of viewing how cases and controls arise in the case-control study and is usually best regarded as a mathematical device for simplifying the calculation of large-sample variances. Another way is for the numbers of cases and controls to be fixed; this is perhaps the most usual assumption. In reality it may be thought unlikely that precise preassigned numbers of cases and controls will be obtained. Usually the numbers used in the study depends on how many individuals become available within the time over which individuals are sampled. A scheme more closely related to the Poisson process is referred to as randomized recruitment, in which each individual in a potential pool of participants is sampled to the case-control study with a specified probability, possibly related to certain features of the individual. This results in the numbers of cases and controls being random rather than fixed. For an example, see Weinberg and Sandler (1991), who used randomized recruitment to a case-control study of the association between radon exposure and lung cancer risk. An alternative way of arriving at a general formulation for case-control studies is a profile likelihood approach described by

Seaman and Richardson (2004). The extension of Bayesian analyses to case-control studies was also outlined in that article.

Section 4.3. Scott and Wild (1986) extended the original argument of Prentice and Pyke (1979) to case-control studies with stratified sampling. Prentice and Pyke (1979) also gave a brief extension to conditional analyses of finely stratified case-control studies, e.g. for use with individually matched studies. The use of conditional analysis based on the prospective model for matched case-control studies is also discussed in Breslow *et al.* (1978).

Section 4.4. Results on the bias which may arise due to the non-collapsibility of odds ratios were given by Gail *et al.* (1984). See Cox and Snell (1989, Section 1.5) for a commentary on the relationship between the logistic and normal distributions.

Section 4.5. The extension of the main result of Prentice and Pyke (1979) to general multiplicative intercept models is discussed by Weinberg and Wacholder (1993), who showed that 'case-control data can be analysed by maximum likelihood as if they had arisen prospectively, up to an unidentifiable multiplicative constant which depends on the relative sampling fractions'. See also Hsieh *et al.* (1985).

Section 4.6. Lynn and McCullogh (1992) considered how adequate the adjustment for covariates needs to be in explaining the matching before a regression adjustment analysis becomes more effective than a matched analysis; they focussed on matched pairs. Neuhäuser and Becher (1997) investigated bias and efficiency from a standard analysis of unmatched case-control study and a conditional analysis based on post-hoc stratification. Armitage (1975) and Siegel and Greenhouse (1973) showed that biased odds ratio estimates are obtained if the matching is ignored and an unmatched analysis used, in a situation where the matching was not random. Feinstein (1987) discussed the effects of ignoring the matching in a matched-pair case-control study in different situations. For a comparable discussion of cohort studies, see De Stavola and Cox (2008). See also Neuhaus and Segal (1993).

Section 4.7. The attributable risk defined in (4.51) is also referred to as the 'population attributable risk', the 'etiologic fraction' and the 'attributable fraction'. It should not be confused with the attributable risk among the exposed individuals, defined for a binary exposure and outcome as $1 - \Pr(Y = 1 | X = 0)/\Pr(Y = 1 | X = 1)$. King and Zeng (2002) suggested a 'robust Bayesian' method of obtaining bounds for the risk ratio and the risk difference by eliciting information from experts about plausible ranges of values for the sampling fractions n_i/N_i. See Kuritz and Landis (1988b) and Benichou and Gail (1990) for estimation of the adjusted attributable risk using the Mantel–Haenszel common odds ratio method, with the inclusion of variance estimators. There is a sizeable literature on methods for estimating risk differences and attributable risks from case-control studies when there is no information available about the number of cases and controls in the underlying population. See, for example, Bruzzi *et al.* (1985), Benichou (1991), Whittemore (1982), Drescher and Schill (1991), Kuritz and Landis (1987, 1988a, b), and Chen (2001).

Section 4.8. An account of the use of information about marginal distributions of the exposure and the outcome is given in Hsieh *et al.* (1985).

5

Case-control studies with more than two outcomes

- Case-control studies can involve more than two outcome groups, enabling us to estimate and compare exposure-outcome associations across groups.
- Studies may involve multiple case subtypes and a single control group, or one case group and two or more control groups, for example.
- Case-control studies with more than two outcome groups can be analysed using pairwise comparisons or special polychotomous analyses. The general formulation based on logistic regression extends to this situation, meaning that the data from such studies can be analysed as though arising from a prospective sample.
- By contrast, in some situations case-only studies are appropriate; in these no controls are required. In one such situation the nature of the exposure contrasts studied may make the absence of controls reasonable. In another, each individual is in a sense his or her own control.

5.1 Preliminaries

In most of this book we are supposing that there are just two possible outcomes for each individual defining them as either cases or as controls. However, there are two contrasting situations where other than two outcomes are involved. First, it may be of interest to estimate and compare risk factors for three or more outcomes; an extended case-control design can be used to make comparisons between more than two outcome groups. The other, very contrasting, situation occurs when controls may be dispensed with, the case-only design. In the first part of this chapter we consider the former situation, and in the second part the latter.

5.2 Polychotomous case-control studies

5.2.1 Preliminaries

Case-control studies with multiple outcomes have been called *polychotomous*, *polytomous* or *multinomial*. In the present context we will use the term *polychotomous*. Such a study may take different forms:

- two or more case groups and a single control group;
- one case group and two or more control groups;
- several case groups and several control groups.

Studies involving more than one case group enable a comparison of the effects of an exposure on different types of case, defined relative to a common control group. For example, in epidemiological studies investigators may wish to compare risk factors for disease subtypes, for example ischaemic and haemorrhagic stroke, oestrogen- and progesterone-receptor breast cancer, adenocarcinoma and squamous cell carcinoma of the lung and aggressive and non-aggressive cancers in the same site. The case groups may be defined prior to case selection and the sampling performed within the relevant subgroups, or they may be defined by subdividing a pre-existing group of cases into subtypes.

Sometimes it may be of interest to compare risk factors for a particular outcome of interest relative to several types of control groups to assess whether risk estimates depend on the choice of control group. For example, one could use hospital controls and population controls or sibling and population controls.

Polychotomous case-control studies may be either matched or unmatched. We will focus on unmatched studies in this section. Matched studies are considered in Section 5.3.

Example 5.1: Case-control study with case subtypes Yaghjyan *et al.* (2011) studied the association between breast density and breast cancer in postmenopausal women, with breast cancer cases divided according to the characteristics of the tumour. The source population was the Nurses' Health Study, which is a prospective cohort of female nurses in the United States. Within this cohort, postmenopausal breast cancer cases were identified during the period from June 1989 to June 2004. Incidence density sampling, as outlined in Section 3.10, was used to select up to two controls for each case at her date of diagnosis. Controls were also matched to cases on age, use of postmenopausal hormones and date and time of blood collection, since the original study was designed to investigate biomarker exposures. The exposure information was obtained from mammogram records. The final analysis was based on 1042 cases and 1794 controls. Information on a range of tumour characteristics was obtained from medical records: invasiveness of the tumour, histological type, grade, size, receptor status and nodal involvement. In this example, case subtypes were identified after

Table 5.1 *Summary of data from a case-control study with three outcomes.*

	$Y = 0$	$Y = 1$	$Y = 2$
$X = 0$	$n_0 - r_0$	$n_1 - r_1$	$n_2 - r_2$
$X = 1$	r_0	r_1	r_2

selection. In a series of analyses, cases were subdivided into two types for each of the above tumour characteristics, for example tumour size ≤ 2 cm or > 2 cm, with the exception of tumour grade, where the cases were divided into three types. Polychotomous unconditional logistic regression, described below in Section 5.2.2, was used to investigate the association of the exposure, breast density, with different case subtypes relative to the controls, after adjustment for potential confounders. Using this approach the authors found evidence that a higher mammographic density is associated with more aggressive tumour characteristics.

5.2.2 Analysis

There are two approaches to analysing data from polychotomous case-control studies.

- In one approach each outcome group is compared with a reference group in a series of separate dichotomous case-control analyses.
- In the alternative approach all outcome groups are studied simultaneously relative to the reference group in a polychotomous analysis.

We begin by considering a polychotomous case-control study with three outcome groups, denoted by $Y = 0, 1, 2$, and a binary exposure X. The data arising from such a study are summarized in Table 5.1.

We proceed as before by considering a *population model* and, derived from that, a *sampling model* and then an *inverse model*.

The inverse model, that is, the formal model for the conditional distribution of Y given $X = x$, takes the form

$$\Pr^{\mathcal{I}}(Y = 0 | X = x) = \frac{1}{1 + e^{\lambda_1 + \beta_1 x} + e^{\lambda_2 + \beta_2 x}},$$

$$\Pr^{\mathcal{I}}(Y = 1 | X = x) = \frac{e^{\lambda_1 + \beta_1 x}}{1 + e^{\lambda_1 + \beta_1 x} + e^{\lambda_2 + \beta_2 x}}, \tag{5.1}$$

$$\Pr^{\mathcal{I}}(Y = 2 | X = x) = \frac{e^{\lambda_2 + \beta_2 x}}{1 + e^{\lambda_1 + \beta_1 x} + e^{\lambda_2 + \beta_2 x}}.$$

Calculation shows that the sampling model, that is, the conditional distribution of X given outcome Y, has the form

$$\Pr^{\mathcal{D}}(X = 1|Y = 0) = \frac{e^{\chi}}{1 + e^{\chi}},$$

$$\Pr^{\mathcal{D}}(X = 1|Y = 1) = \frac{e^{\chi+\beta_1}}{1 + e^{\chi+\beta_1}}, \qquad (5.2)$$

$$\Pr^{\mathcal{D}}(X = 1|Y = 2) = \frac{e^{\chi+\beta_2}}{1 + e^{\chi+\beta_2}},$$

where χ is a relatively complicated function of the other parameters whose form we do not need. Note that, for example,

$$\log \frac{\Pr(X = 1|Y = 1)\Pr(X = 0|Y = 0)}{\Pr(X = 1|Y = 0)\Pr(X = 0|Y = 1)} = \beta_1. \qquad (5.3)$$

Hence β_1 is the log of the ratio of the odds of exposure in the $Y = 1$ group and the odds of exposure in the $Y = 0$ group. There are similar interpretations for β_2 and for $\beta_2 - \beta_1$.

The sampling model yields the log likelihood

$$l_{\mathcal{D}} = r.\chi + r_1\beta_1 + r_2\beta_2 - n_0 \log(1 + e^{\chi}) - n_1 \log(1 + e^{\chi+\beta_1})$$
$$- n_2 \log(1 + e^{\chi+\beta_2}), \qquad (5.4)$$

whereas the inverse model yields the log likelihood

$$l_{\mathcal{I}} = n_1\lambda_1 + n_2\lambda_2 + r_1\beta_1 + r_2\beta_2 - (n - r.) \log(1 + e^{\lambda_1} + e^{\lambda_2})$$
$$- r. \log(1 + e^{\lambda_1+\beta_1} + e^{\lambda_1+\beta_2}), \qquad (5.5)$$

where $r. = r_0 + r_1 + r_2$.

It follows, most directly from the sampling model, that the overall model and the three separate two-outcome models obtained by omitting one set of outcomes at a time yield identical estimates of the parameters of interest, namely β_1 and β_2. This shows that the same estimates of β_1 or of β_2 will be obtained from dichotomous analyses that compare the $Y = 1$ group or the $Y = 2$ with the $Y = 0$ group and from a polychotomous analysis in which the three groups are modelled simultaneously. This would not be the case, however, if additional terms representing confounding effects were included, with parameters common to the three possible outcomes. Any extension of the model that does not preserve the symmetry will result in different estimates.

Although the same estimates of β_1 and β_2 can be obtained from two dichotomous analyses as from a polychotomous analysis, the apparent precision of the estimates may differ. This is considered further in Section 5.2.4.

5.2.3 A general formulation

The general formulation for the analysis of case-control data, described in Chapter 4, extends to studies with more than two outcome groups, enabling analysis involving continuous exposures and regression adjustment for other variables.

We define binary variables Y_d ($d = 0, 1, \ldots, D$), where Y_d takes the value 1 for an individual in outcome group d and the value 0 otherwise. The group with $Y_0 = 1$ is defined as the reference or *control group*. Those with outcomes $d = 1, \ldots, D$ are referred to as the *case* groups, although this particular interpretation is not necessary. We consider a vector of exposures X.

The inverse model is the conditional probability that an individual is in outcome group d ($d = 0, 1, \ldots, D$), given that they are in one of the possible $D + 1$ outcome groups, and is specified by

$$\Pr^{\mathcal{I}}(Y_d = 1 | X = x) = \frac{\exp\left(\alpha_d + \beta_d^T x\right)}{\sum_{d=0}^{D} \exp\left(\alpha_d + \beta_d^T x\right)}, \tag{5.6}$$

where $\alpha_0 = \beta_0 = 0$ and β_d is a vector of parameters that have interpretations as log odds ratios for outcome group d relative to the baseline group with $Y = 0$.

Polychotomous case-control data can be analysed as though the data arose prospectively using a likelihood that is the product over all individuals of the factors

$$\Pr^{\mathcal{D}}(Y_d = 1 | X = x) = \frac{\exp\left(\alpha_d^* + \beta_d^T x\right)}{\sum_{d=0}^{D} \exp\left(\alpha_d^* + \beta_d^T x\right)}, \tag{5.7}$$

where $\alpha_d^* = \alpha_d + \log(\eta_d/\eta_0)$ and where η_d, with $d = 0, 1, \ldots, D$, is the ratio of the probability of outcome d in the polychotomous case-control sample and the probability in the population. This is analogous to the result given in Section 4.2 for a single case type.

For a pairwise analysis of polychotomous case-control data each case group is compared separately with the control group in a series of separate dichotomous regressions. The relevant conditional probabilities from which the likelihood is formed are

$$\mathrm{Pr}^{\mathcal{D}}(Y_d = 1 | X = x, Y_d + Y_0 = 1) = \frac{\exp\left(\alpha_d^* + \beta_d^T x\right)}{1 + \exp\left(\alpha_d^* + \beta_d^T x\right)}. \tag{5.8}$$

5.2.4 Pairwise versus polychotomous analyses

In an analysis in which all outcome groups are considered simultaneously using (5.7), an estimate of the asymptotic covariance matrix, V say, for the estimates $\tilde{\beta} = (\tilde{\beta}_1, \ldots, \tilde{\beta}_D)^T$ is obtained directly from the maximum likelihood analysis, since all outcomes are treated in a single log likelihood. The matrix V is estimated by the inverse of the appropriate information matrix. This follows from an extension of the results in Chapter 4. Performing tests of the equality of exposure effects on different case groups is therefore straightforward using the results from this polychotomous analysis.

An analysis in which outcome levels are taken a pair at a time is in a sense more complicated, in that the covariance matrix associated with two different pairs having a common comparison group has to be considered. For example, if we analyse the pairs of outcomes with $d = 0, 1$ and with $d = 0, 2$ separately, a comparison of the parameters associated with $d = 1, 2$, that is, a comparison of β_1 and β_2, is direct but the standard errors involved need to take account of the dependence induced by the common data associated with $d = 0$. In such a pairwise analysis, obtaining the covariance matrix of the estimates requires study of the column vector formed by stacking the *score vectors* $U_d(\beta_d)$ with components $\partial l_d / \partial \beta_{dj}$. The corresponding full information matrix $I(\beta)$ is block diagonal with information matrices from separate pairwise analyses as blocks. The covariance matrix is then found using the sandwich formula; see the appendix.

There is a potential gain in efficiency from the polychotomous analysis because the comparison between the case groups, as well as between each case group and the control group, in a polychotomous analysis provides additional information about the log odds ratio parameters. We omit the details, which are, however, needed for a complete assessment of the efficiency of the alternative analyses. Modern statistical software packages allow for polychotomous logistic regression; hence there is no computational advantage in using the pairwise approach.

5.3 Polychotomous studies with matching

We now concentrate on matched case-control studies involving polychotomous outcomes. Suppose first that the focus is on the comparison of two or more case groups with a control group. Matched polychotomous case-control data may be obtained in one of two main ways.

(1) If two or more case groups are defined prior to the selection of individuals for the case-control sample then an attempt may be made to form matched case-control sets containing at least one individual in each outcome group, including the control group.
(2) Another possibility is that a matched case-control study is formed first using a single case type. As a secondary analysis it may be of interest to divide the cases into subtypes, for example disease subtypes, as noted in Section 5.2.1. In this situation the matched sets will contain only one type of case and the controls.

In the second situation, standard analyses for matched case-control data could be applied by separating the matched sets into those containing different case subtypes. The matched sets could be combined into one analysis if the effects of certain covariates were assumed to be the same across case types. This would have advantages in terms of formal efficiency.

The first situation is more interesting to consider, and there are two possible analyses. In one, a polychotomous analysis that uses all members of each matched set is performed. In the other, pairwise analyses are performed; in these analyses each case type is considered with its matched controls, the other case types in each matched set being ignored. We consider the two possibilities in more detail.

If m denotes features common to individuals in the mth matched set, the inverse model of interest is

$$\Pr^{\mathcal{I}}(Y_d = 1 | X = x, M = m) = \frac{\exp(\alpha_{dm} + \beta_d x)}{\sum_{d=0}^{D} \exp\left(\alpha_{dm} + \beta_d^T x\right)}. \qquad (5.9)$$

After conditioning on selection to the case-control sample, it is modified to

$$\Pr^{\mathcal{D}}(Y_d = 1 | X = x, M = m) = \frac{\exp\left(\alpha_{dm}^* + \beta_d x\right)}{\sum_{d=0}^{D} \exp\left(\alpha_{dm}^* + \beta_d^T x\right)}, \qquad (5.10)$$

where $\alpha_{dm}^* = \alpha_{dm} + \log\{\eta_d(m)/\eta_0(m)\}$ and where $\eta_d(m)$ for $d = 0, 1, \ldots, D$ is as before the ratio of the probabilities of outcome group d in the data and in the population, respectively, for the mth matched set.

We consider specifically the situation in which each matched set is a triplet comprising two case types and one control, with binary exposure. The exposure for the individual in outcome group $d = 0, 1, 2$ in the uth matched triplet is denoted $X_{u;y}$. Therefore, for example in a matched triplet u with two exposed members and one unexposed, the conditional probability that the two exposed members are cases is given by

$$\Pr(X_{u;0} = 0, X_{u;1} = 1, X_{u;2} = 1 | X_{u;0} + X_{u;1} + X_{u;2} = 2)$$

$$= \frac{e^{\beta_1 + \beta_2}}{e^{\beta_1 + \beta_2} + e^{\beta_1} + e^{\beta_2}}. \tag{5.11}$$

Let n_{abc} denote the number of matched sets in which individuals with outcomes 0, 1, 2 have respectively exposures a, b, c. The full conditional likelihood is

$$\beta_1(n_{011} + n_{110} + n_{010}) + \beta_2(n_{011} + n_{101} + n_{001})$$
$$- (n_{011} + n_{110} + n_{101}) \log(e^{\beta_1} + e^{\beta_2} + e^{\beta_1 + \beta_2})$$
$$- (n_{100} + n_{010} + n_{001}) \log(1 + e^{\beta_1} + e^{\beta_2}). \tag{5.12}$$

If instead we were to focus on a pairwise comparison between outcome groups $d = 0, 1$, ignoring information on the individual with $d = 2$ in each matched set, the conditional likelihood used would be

$$n_{01.}\beta_1 - (n_{01.} + n_{10.}) \log(1 + e^{\beta_1}); \tag{5.13}$$

this pairwise or dichotomous comparison of groups $d = 0, 1$ uses only discordant pairs, that is, it uses a total of $n_{10.} + n_{01.}$ matched sets, where $n_{ab.} = n_{ab0} + n_{ab1}$. The polychotomous analysis, however, discards only the $n_{000} + n_{111}$ fully discordant sets. To estimate β_1 the polychotomous analysis can therefore utilize $n_{10.} + n_{01.} + n_{110} + n_{001}$ sets, more of the data than the corresponding pairwise analysis.

Thus polychotomous and dichotomous treatments of matched polychotomous data use somewhat different parts of the data. Therefore in general the two approaches will yield different estimates of the parameters of interest. This is in contrast with the unmatched situation, where dichotomous and polychotomous analyses yield the same estimates albeit with different apparent precision.

The conditional likelihood for a general polychotomous matched case-control study is

$$\prod_u \frac{\exp\left(\sum_{d=0}^{D} \beta_d \sum_{i=1}^{n_{u;d}} x_{u;di}\right)}{\sum_{p \in P_{\xi_u}} \exp\left(\sum_{d=0}^{D} \beta_d \sum_{l \in \xi_u^{p_d}} x_l\right)}, \tag{5.14}$$

where $x_{u;di}$ denotes the exposure for the ith individual with outcome d in the uth matched set; $n_{u;d}$ denotes the number of individuals with outcome d ($d = 0, 1, \ldots, D$) in the uth matched set; ξ_u denotes the set of all individuals in the uth matched set; P_{ξ_u} is the set of possible partitions of the set ξ_u into $D + 1$ subsets of sizes $n_{u;0}, n_{u;1}, \ldots, n_{u;D}$; and $\xi_u^{P_d}$ is the subset of individuals assigned to group d in the pth partition of the set ξ_u. The polychotomous analysis can be extended to allow for incomplete matched sets, that is, matched sets in which not all of the outcomes appear.

5.4 More than one control group

As already noted, sometimes more than one control group is selected for a given case group. The different control types used may include population controls, hospital controls, neighbourhood controls, or family controls. The reasons for selecting more than one control group include the following.

- Obtaining similar exposure-outcome association estimates using different control groups can provide stronger evidence for an association, because it suggests that the association is not due to any bias induced by control selection.
- If there are differences in exposure-outcome associations using different control groups then the nature of these control groups may help to explain where the differences arise. For example, different associations obtained using sibling and population controls could suggest the presence of an important genetic component.

The issues surrounding the analysis of studies with more than one control group are similar to those discussed in the above sections, where we placed more emphasis on multiple case groups. It will often be of particular interest to perform a test of homogeneity of the estimates obtained for the association between an exposure and being a case using different control groups. As above, the test is direct if a polychotomous analysis is used. If instead it is based on a series of pairwise analyses then the test will underestimate the variance of the differences between estimates unless the correlation between the estimates is estimated separately.

There arise different difficulties for analysis if one control group is matched to the cases and the other is unmatched (for example, the matched controls could be siblings while the unmatched controls are drawn from the underlying population). One option is to break the matching for the matched control group and perform a polychotomous unmatched analysis, with adjustment for matching variables as far as is possible. An alternative approach which retains the matching in the matched part of the study, is possible.

We consider the combination of matched and unmatched studies in more detail, by referring to a study comprising three types of individual: cases ($Y = 1$), a set of unmatched controls ($Y = 0_U$) and a set of controls, each of which is matched to a case on variables denoted M ($Y = 0_M$).

In the unmatched part of the study we assume the model

$$\Pr^{\mathcal{D}_U}(Y = 1 | X = x, W = w) = L_0(\alpha_1^* + \beta_1^T x + \gamma_1^T w), \qquad (5.15)$$

where the superscript $\mathcal{D}_U$ denotes the sampling model for the *unmatched* sample. Here X denotes a vector of the main exposures and W a vector of adjustment variables. In the matched part of the study we assume a model

$$\Pr^{\mathcal{D}_M}(Y = 1 | X = x, M = m, W^* = w^*) = L_0(\alpha_{1m}^* + \tilde{\beta}_1^T x + \tilde{\gamma}_1^T w^*), \qquad (5.16)$$

where the superscript $\mathcal{D}_M$ denotes the sampling model for an individual's being in the *matched* sample. Here W^* denotes a set of adjustment variables not accounted for by the matching. The matching variables, summarized by M, may include only variables in W or parts of W plus additional features that are not part of W.

The models (5.15) and (5.16) are in general incompatible; that is, they cannot both be the true model. This is so, even if all the variables W are incorporated into M and vice versa, owing to the assumptions regarding the functional forms for the adjustment variables in the models. This is also related to non-collapsibility. Perhaps the most likely scenario is that the matched part of the study will adjust, via the matching, for variables that cannot be adjusted for in the unmatched part of the study. A large difference between β and $\tilde{\beta}$ would suggest some element of the matching to be a strong confounder of the exposure-outcome association. However, for smaller differences one cannot completely rule out the possibility that a difference in the association has arisen from other issues related to the inconsistency between the two models.

5.5 Ordered outcome groups

The methods discussed so far in the current chapter are appropriate for studies in which the outcome groups are purely nominal. However, there are circumstances in which the outcome groups have a natural ordering, for example where case groups correspond to disease states of increasing severity. Ordinal disease states can, of course, be treated as nominal and the data analysed using the methods described in previous sections. However, there are also a number of ways in which the ordinal character of the states can be incorporated into the analysis. Here we give only an outline treatment.

One possibility is to obtain separate estimates of, say, particular effects $\hat{\beta}_1, \ldots, \hat{\beta}_d$ taking no account of the ordinal character of d. Then, whenever the estimates do not satisfy the ordinal property, estimates breaking that order are modified by a procedure evocatively called *pool-adjacent violators* until they do. For example, if in a sequence that was broadly increasing one found

$$\hat{\beta}_1 < \hat{\beta}_2 < \hat{\beta}_4 < \cdots$$

but also

$$\hat{\beta}_1 < \hat{\beta}_3 < \hat{\beta}_2$$

then one would pool $\hat{\beta}_2$ and $\hat{\beta}_3$. In more complicated cases the pooling would continue until monotonicity is achieved. Of course, a need for too much pooling might throw suspicion on the assumed monotonicity.

A second possibility is to treat the outcome group number d as if it were a meaningful quantitative variable. This might, for example, suggest a simple representation such as $\beta_d = \gamma_0 + \gamma_1 d$. This may at first seem in contradiction with the distinction between ordinal and quantitative features. However, good ordinal scales are defined in such a way that the gaps between levels are in some sense meaningful, and a linear representation may, subject especially to checking of the role of the extreme points, be an adequate representation.

Formally, in some ways the most satisfactory method for preserving the arbitrary ordinal character of d is to postulate a steady progression of dependence between levels of the ordinal variable. In the simplest version of this, the effect of exposure on the cumulative log odds is constant. That is, for a binary exposure,

$$\frac{\Pr(Y_i \geq d | X_i = 1)/\Pr(Y_i < d | X_i = 1)}{\Pr(Y_i \geq d | X_i = 0)/\Pr(Y_i < d | X_i = 0)} = \Omega \qquad (5.17)$$

for all levels d. This representation has an appealing interpretation in terms of a latent variable arbitrarily partitioned to form ranked outcomes. Nevertheless, the representation imposes a constraint on the relations and as such is not always preferable to treating the rank order as a quantitative outcome.

In all analysis of such data, care is needed to ensure that parameters estimated from the inverse model have the right interpretation in the underlying population or sampling model. This typically involves establishing some relation with odds ratios in 2×2 tables. Thus the log odds

$$\log \frac{\Pr(Y \geq d | X = x)}{\Pr(Y < d | X = x)} \tag{5.18}$$

may be estimated as a function of unknown parameters for each d. The efficient combination of parameters combining analyses for different d requires a consideration of the covariances between estimates referring to different d and the use of generalized least squares.

A commonly used model for ordinal outcomes is the proportional odds model, in which the prospective probability of an individual's being in outcome group d or higher is modelled as

$$\Pr(Y \geq d | X = x) = \frac{e^{\lambda_d + \beta x}}{1 + e^{\lambda_d + \beta x}}. \tag{5.19}$$

The likelihood in a prospective analysis would be the product of terms $\Pr(Y_i = d | X_i = x_i)$, where

$$\Pr(Y = d | X = x) = \Pr(Y \geq d | X = x) - \Pr(Y \geq d + 1 | X = x).$$

In a case-control study the analysis must take into account retrospective sampling. Under the proportional odds model the probability $\Pr(Y_i = d | X_i = x_i)$ is not of multiplicative intercept form and hence the results of Chapter 4 do not apply. Estimates of selection probabilities are therefore required for the use of proportional odds models in case-control studies. We refer to the papers listed in the notes at the end of the chapter for details of the resulting analyses for unmatched and matched studies.

5.6 Case-only studies

5.6.1 Preliminaries

Two central features of the topics discussed in this book are the comparison of cases with controls and the retrospective nature of the data. We now discuss briefly comparable problems *without* a specific set of control individuals. This can be relevant in two different situations:

- in one situation, the nature of the exposure contrasts studied may make the absence of controls reasonable;
- in another, each individual is in a sense his or her own control.

5.6.2 Study of specific effects

It is simplest to discuss the first situation in a special context, that of the interaction between two binary exposures, although the argument is quite general. Suppose that there is an exposure vector X and that analysis of an unmatched case-control study is proposed in terms of a logistic model saturated with parameters for X, there being no confounding variable W in the model. For example, suppose that X has two components X_1 and X_2, both binary, and that an analysis is proposed in terms of the main effects of X_1 and X_2 and their interaction. The inverse model of interest is

$$\Pr{}^{\mathcal{I}}(Y = y | X_1 = x_1, X_2 = x_2) = L_y(\alpha + \beta_1 x_1 + \beta_2 x_2 + \beta_3 x_1 x_2). \quad (5.20)$$

This gives three parameters to describe the three independent contrasts possible among the four levels of the two factors. In the unmatched study let $(r_{ij:1}, r_{ij:0})$ denote the numbers of cases and controls respectively having $X_1 = i$, $X_2 = j$. Then the relevant parameters are estimated from analogous contrasts of the log frequencies; in particular, the interaction is estimated by

$$\log \frac{(r_{11:1} r_{00:1})}{(r_{10:1} r_{01:1})} - \log \frac{(r_{11:0} r_{00:0})}{(r_{10:0} r_{01:0})}. \quad (5.21)$$

By the argument of Section 2.3, the asymptotic variance of (5.21) is the sum of the reciprocals of the eight frequencies in the expression. Now suppose that cases are very rare, so that the distribution of exposures in the controls is essentially that in the originating population and in that population X_1 and X_2 may be assumed to be independently distributed. Then the second term in (5.21) differs from zero only by random fluctuations and the interaction is best estimated by the first term, that is, from data on the cases only. Also, the variance of the resulting estimate is estimated by the sum of four reciprocal frequencies, in contrast with the sum of eight such terms in the original estimate. In summary, not only can the interaction be estimated without controls but it is more statistically efficient to do so, provided, of course, that the primary assumption of the independence of (X_1, X_2) in the population is a reasonable approximation.

Example 5.2: Case-only studies in genetic epidemiology The primary application of case-only studies is in genetic epidemiology, in the study of gene-environment interaction. In its simplest version we have two binary exposure variables, X_1, representing an environmental factor, and X_2, representing a biallelic gene. We define $X_3 = X_1 X_2$ to capture an interactive effect. More specifically, the coefficient of X_3 captures the consequences of, say, a rare genetic form combined with an extreme environmental exposure. If in the population the genetic and environmental factors may be assumed independently distributed, a case-only study is indicated. A possible compromise is to include a limited number of controls to provide some security against major failure of the critical-independence assumption.

5.6.3 *Using cases as self-controls*

In some situations it can be advantageous for observed cases to indirectly provide their own control information. There are different ways in which this can work, which we discuss below.

Consider an exposure of interest that is transient, that is, it occurs at an instant or over a short time period. We wish to study the possible effects of the exposure when it may be assumed that it can have a direct consequence only for a short time period after its occurrence. In this type of study the event of interest is an acute event, such as an adverse reaction or an emergency hospital admission. Examples of a transient exposure are the following.

- *A vaccine*: we might be interested in potential adverse effects in the days following receipt of the vaccine.
- *A peak in the level of air pollution*: this may be studied in relation to hospital admission for respiratory conditions.

In the discussion below we focus on binary exposures for simplicity but the methods are not restricted to this situation.

Now suppose that cases of an acute event of interest have been observed and that the exposure status of each individual at the time of the event observed for him or her is available. For these individuals there will have been times in the past when the acute event might have occurred. For each case, therefore, we could define one or more comparable occasions in the past when the individual did not have the acute event. These form 'control occasions' for each 'case occasion'. The exposure status of the individual on the control occasions is also required and is established retrospectively from the individual, or, perhaps, using relevant records. For each individual the case occasion and the control occasions form a matched set and the data may be analysed using conditional logistic regression. See Figure 5.1.

This type of design is sometimes referred to as a *case-crossover study*. There is an analogy with crossover experiments comparing two treatments,

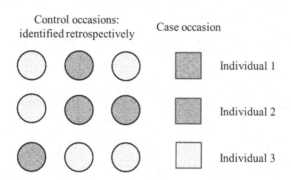

Control occasions: identified retrospectively

Case occasion

Individual 1

Individual 2

Individual 3

Figure 5.1 A study using self-controls: each case (square) provides one or more control occasions or periods (circles) from the past. On each occasion it is observed retrospectively whether the individual was exposed (mid-grey) or unexposed (pale grey), resulting in a matched case-self-control study.

in which all individuals in the study receive both treatments in succession, typically with a 'wash-out' period in between.

Example 5.3: Case-crossover study Redelmeier and Tibshirani (1997) studied minor traffic accidents in Toronto. They had access to the billing records of mobile phone companies and so were able to determine in an approximate fashion whether a driver involved in an accident was using a mobile phone at the time. They then took as a control period the same time one week previously (and a second control period two weeks previously). Unfortunately they could not assess directly whether the individual was driving in the previous period but estimated this by an independent study. Thus, taking the binary exposure as the use of a mobile phone there was reasonably reliable information on the exposure of the cases (accidents) but only a probabilistic estimate of the exposure of each control. The form of the data is otherwise that of a matched case-control study with controls formed from the previous history of a case.

A different use of self-controls is not to select comparable control occasions for each case, as above, but instead to obtain information on the length of exposure over a given time period. This situation, illustrated in Figure 5.2, has been referred to as the *self-controlled case-series method*. Suppose now that each individual is studied over a time interval, say (a_i, b_i) for individual i. In this period there are one or more instants or periods of exposure to the risk under study or its direct consequences. The remaining interval is divided into 'dead' periods, in which either no exposure was

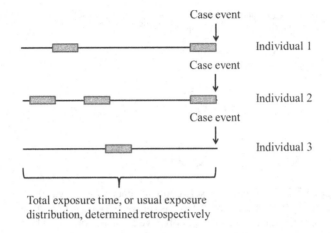

Total exposure time, or usual exposure
distribution, determined retrospectively

Figure 5.2 Another study using self-controls: for each case,
information is available on the total exposure time (shaded
blocks) over a particular period, for example the month prior to an
individual's becoming a case. Alternatively, information may be
obtained for each case on the case's 'usual' frequency of exposure.

experienced or any direct effect of exposure may safely be assumed to be
negligible, and 'neutral' periods whose status is ambiguous. The neutral
periods may be omitted. During the whole period of observation, critical
events may occur and may be regarded as defining cases. The objective is
to study any possible effect of exposure on the critical events. In a retrospective setting we may observe just one event (the case occasion), which
may be during a period of exposure or non-exposure, and then look back
over a fixed period of time to assess the total time the individual spent in
an exposure state. Alternatively, it might be possible to elicit information
about the 'usual' frequency of exposure.

In the following discussion we will assume that, in the absence of exposure, cases occur in Poisson processes at rates specific to individuals. In
many epidemiological applications in which observations on an individual
extend over an appreciable time span it will be desirable to introduce an age
dependence into the analysis, typically by defining a number of age bands
each with its specific contribution to the rate of occurrence of cases. Here
we omit that complication.

The formal analysis proceeds as follows. Suppose that adverse events on
individual i occur in a Poisson process with rate

$$\exp\{\lambda_i + \beta\tilde{x}(t)\}, \tag{5.22}$$

where t denotes time, $\tilde{x}(t) = 0$ in a dead period and $\tilde{x}(t) = 1$ in an exposed period. The nuisance parameter λ_i specified to the ith individual is eliminated from the likelihood by conditioning on the total number n_i of events experienced by that individual. This number has a Poisson distribution, and the resulting conditional likelihood is that of a binomial distribution. Let n_{i1}, n_{i0} be the numbers of events in the exposed and dead periods respectively and let x_{i1}, x_{i0} be the corresponding exposure periods, namely

$$x_{i1} = \int_{a_i}^{b_i} \tilde{x}(t)dt, \quad x_{i0} = b_i - a_i - x_{i1}. \tag{5.23}$$

Then the conditional likelihood is proportional to

$$\prod_i \left(\frac{x_{i1}e^{\beta}}{x_{i0} + x_{i1}e^{\beta}} \right)^{n_{i1}} \left(\frac{x_{i0}}{x_{i0} + x_{i1}e^{\beta}} \right)^{n_{i0}}. \tag{5.24}$$

Additional effects may be included in the model either to eliminate possible time trends or to allow for interactive effects. The above analysis using Poisson processes relies on knowledge about the amount of exposure time over a given time period.

In the above situation a hazard-based analysis may also be possible. In this we assume that each individual i has a hazard, $h_i(t)$ say, for the event of interest. One possible form is the proportional hazards representation, for which

$$h_i(t) = h_{0i}(t)e^{\beta x_i(t)}, \tag{5.25}$$

where $h_{0i}(t)$ is a baseline hazard, that is, the hazard under no exposure, and where $x_i(t)$ denotes the exposure status of individual i at time t. Now let $p_i(t)$ denote the marginal probability of exposure at time t for individual i. Under the above hazard model the distribution of the exposure, conditional on individual i having the event at t, can be written as

$$\frac{e^{\beta x_i(t)} p_i(t)}{e^{\beta x_i(t)} p_i(t) + 1 - p_i(t)}. \tag{5.26}$$

The likelihood is the product of the terms in (5.26) over all individuals who had the event. An analysis therefore requires information about $p_i(t)$.

As noted at the start of this section, studies using self-controls are suitable for use when the exposure of interest is transient and the outcome acute. The benefit of using a case to provide effectively his or her own control is that any bias due to systematic differences between the case group and the control group that might arise in a standard case-control study using distinct individuals as controls is eliminated. For example, in the traffic accident study described in Example 5.3 it might have been difficult to choose controls for each case that matched the case on 'driving performance'. One could attempt matching on variables such as years of driving experience, frequency of car use, type of vehicle and so on, but it might be difficult to capture the most important factors. The case itself provides a very appropriate control. There are some issues to consider when using self-controls, however. It is possible that the effects of the exposure may differ at different time points. In the first design described above (Figure 5.1) this could be alleviated in some situations by careful choice of the control occasions. There is also the possibility of within-person confounding; this occurs if some other transient exposure(s) affects the main transient exposure. In the situation of Figure 5.1, adjustment for this is possible provided both exposures are observed on the case occasion and on each control occasion. Similarly, both exposures may be observed under the Poisson process representation. However, in a situation where the normal frequency of exposure is used it could be difficult to ascertain this sort of information.

Example 5.4: Self-controlled case-series study   Whitaker *et al.* (2009) described the use of the self-controlled case series approach in a study of an adverse event following measles, mumps and rubella (MMR) vaccinination in the UK. The adverse event was hospital admission for a rare autoimmune disorder. The six-week period after MMR vaccination was divided into three two-week risk periods. Cases of the adverse event were observed in a three-year period among children aged 12–23 months, and the timing of the events relative to the timing of the MMR vaccination was investigated. The relative incidence of the adverse event was found to be significantly higher in the second two-week period following receipt of the vaccine.

5.6.4 Connection with specific stochastic process models

There is a broad connection of the analysis of self-controlled case-series with the analysis of systems representable by stochastic process models, in particular by point processes. Data from such systems have two broad forms:

- They may consist of a single, or a small number of, long series within each of which there is appreciable internal replication.
- Alternatively, as in the previous discussion, there are many short series, obtained from independent individuals.

There are many possible forms of stochastic model that may arise for such situations, for example point process models with two types of point, one type explanatory and the other representing an outcome. A central question in all such studies is how to achieve a reasonably precise estimation of the effect of an explanatory feature, an estimation that is free of contaminating influences and allows for individual differences.

We discuss briefly one such situation arising in an industrial environment. A machine producing a long sequence of items has an intermittent fault, which is immediately repaired but has a possible adverse outcome over a short time interval. Such events constitute a point process of explanatory events, of type X, say. The output sequence of items produced has occasional defects, point events of type Y, say. It may be required to assess the possible effect of X in producing the outcomes Y. It is supposed initially that any effect of a type-X event at time t' is confined to the fairly short time interval $(t' - h_0, t' + h_1)$. Here (h_0, h_1) are supposed given *a priori*, although in an extended version of the problem they may be estimated.

In the following discussion we suppose, without essential loss of generality, that $h_0 = 0$ and write $h_1 = h$. In a very crude analysis we would define a time interval of length h following each X-event. Suppose that the union of these is of length $m(h)$, the total period of observation being of length t_0. Then if n_Y events of type Y are observed in total, the random variable, $N_Y(h)$, representing the type-Y events falling in $m(h)$ will have, under the simplest of assumptions, that is, the null hypothesis of no dependence, a binomial distribution of index n_Y and parameter $m(h)/t_0$.

Under a non-null model, suppose that each type-X event generates a type-Y event with probability θ. Then θ may be estimated by the following argument. The type-Y events are of two sub-types. First, $n_y - n_y(h)$ are observed outside the period $m(h)$ and therefore are not associated with a type-X event. That is, such events occur at an estimated rate $\{n_y - n_y(h)\}/\{t_0 - m(h)\}$. Therefore the estimated number of such events occurring in the period $m(h)$ is

$$\frac{m(h)\{n_y - n_y(h)\}}{t_0 - m(h)}. \tag{5.27}$$

The remaining type-Y events in the period $m(h)$ provide an estimate of those associated with a type-X event. This number divided by n_x is thus an estimate of θ, namely

$$\frac{n_y(h)t_0 - n_y m(h)}{n_x\{t_0 - m(h)\}}, \tag{5.28}$$

where n_x is the observed number of type-X events. The estimate would be constrained to the interval $(0, 1)$.

The statistical properties of this estimate are to be considered conditionally on the realized type-X points. The estimate is, however, sensitive to non-randomness, for example to common trends in both types of event.

Application of the matched case-control method to such a situation would involve defining both before and after each type-X event, and reasonably close to it, a pseudo event and repeating the estimation procedure on these. Use of virtually all allowable time points as controls would be close to the use of (5.28). Alternatively, some control over trends would be achieved by stratifying the data into time periods, estimating θ separately from each stratum and combining the estimates only if mutually consistent.

Notes

Section 5.2.1. The comparison of three or more outcome groups in a polychotomous case-control study is the case-control equivalent to an experimental study in which there are more than two possible responses to a treatment. Polychotomous and pairwise logistic analyses for a polychotomous dose-response situation were described by Mantel (1966).

Section 5.2.2. An analysis of unmatched polychotomous case-control studies was described by Begg and Gray (1984). See also Dubin and Pasternack (1986), Bull and Donner (1993) and Marshall and Chisholm (1985). Thomas *et al.* (1986) described the analysis of polychotomous studies with multiple case groups but no control group, referred to as case-heterogeneity studies.

Section 5.2.4. Begg and Gray (1984) presented efficiency calculations for the polychotomous versus the pairwise analysis of polychotomous case-control data with three outcome groups and two binary exposures. The asymptotic relative efficiency for individual parameter estimates was obtained in a range of situations and for joint tests of the effect of an exposure on different outcome groups. Similar efficiency calculations for normally distributed exposures were performed by Bull and Donner (1993). Whittemore and Halpern (1989) devised an alternative score test for homogeneity across two or more disease categories in a polychotomous matched case-control study using the polychotomous logistic model. To perform the test the polychotomous logistic model is reparameterized by writing $\beta_d = \beta^* + \gamma_d$ $(d = 1, \ldots, D)$, and β^* is eliminated in a conditional analysis. This test requires less computation than the standard test, because it excludes the control covariates, and it is asymptotically equivalent to the standard test.

Section 5.2.3. Prentice and Pyke (1979) incorporated the possibility of multiple types of outcome in their derivation of a general formulation for the analysis of case-control data, as outlined in Chapter 4 for the standard situation of two outcome groups.

Section 5.3. Matched studies have received less attention but have been described by Liang and Stewart (1987), Levin (1988), Risch and Tibshirani (1988), Whittemore and Halpern (1989) and Becher (1991).

Section 5.4. The combining of case-control studies with matched or unmatched controls has been considered by Le Cessie *et al.* (2008) and Gebregziabher *et al.* (2010).

Section 5.5. For a general discussion of models for ordinal response data see, for example, Agresti (1984, 1990) and McCullagh (1980). For the pool-adjacent violators algorithm, see Barlow *et al.* (1972). See Liu and Agresti (1996) for a discussion of a combined estimate for cumulative odds ratios and a variance estimate. The estimation of a cumulative odds ratio from case-control studies when the selection probabilities are known is related to the work of Wild (1991) and Scott and Wild (1991, 1997) on the analysis of case-control data supplemented by population totals for each outcome group. Wild (1991) compared the efficiencies of weighted versus conditional pseudo-likelihood estimation for a proportional odds model for multiple ordinal outcomes in a case-control study. Mukherjee *et al.* (2008) discussed stratified proportional odds models in the context of prospective data. A combined Mantel–Haenszel estimate for stratified matched case-control studies with ordinal outcomes and its variance were derived by Liu and Agresti (1996). Mukherjee *et al.* (2007) also discussed the analysis of matched studies with ordinal outcomes, including the use of Bayesian methods and random effects models.

Section 5.6.2. Piegorsch *et al.* (1994) discussed case-only designs for estimating interactions. See Khoury and Flanders (1996), Gatto *et al.* (2004) and VanderWeele *et al.* (2010) for the use of case-control studies in investigations of gene-environment interactions.

Section 5.6.3. The case-crossover design was described by Maclure (1991). For an overview of the different variations on this design see Mittleman *et al.* (1995). Note that Maclure (1991) used a simple Mantel–Haenszel estimate for the event rate, which we do not discuss here. The self-controlled case-series method was introduced by Farrington (1995). Important theoretical work and relevant applications were reviewed by Whitaker *et al.* (2006, 2009).

Section 5.6.4. The industrial example discussed in this subsection is based on Cox and Lewis (1966).

6

Special sampling designs

- In two-stage case-control designs, limited information is obtained on individuals in a first-stage sample and used in the sampling of individuals at the second stage, where full information on exposures and other variables is obtained. The first stage may be a random sample or a case-control sample; the second stage is a case-control sample, possibly within strata. The major aim of these designs is to gain efficiency.
- Two-stage studies can be analysed using likelihood-based arguments that extend the general formulation based on logistic regression.
- Special sampling designs for matched case-control studies include counter-matching, which uses some information on individuals in the potential pool of controls to select controls in such a way as to maximize the informativeness of the case-control sets.
- Family groupings can be used in case-control-type studies, and there is a growing literature in the epidemiological, statistical and genetics fields. In one approach, cases are matched to a sibling or other relative.

6.1 Preliminaries

So far we have discussed case-control studies in which cases and controls are sampled, in principle at random, from the underlying population on the basis of their outcome status. We have also considered extensions, including matched studies and stratified sampling, in both of which it is assumed that some features of individuals in the underlying population are easily ascertained. Sometimes it is useful to consider alternative ways of sampling in a case-control study. In this chapter we discuss some special case-control sampling designs.

We will devote much of this chapter to a discussion of two-stage sampling designs. In these, limited information is obtained on individuals in a first-stage sample and used in implementing the sampling of individuals at the second stage. The major aim of two-stage sampling is to gain efficiency, that is, to obtain more precise estimates of the primary relationships of interest.

We use the term *two-stage*, but in the survey sampling literature the term *two-phase* is commonly used. The analysis of two-stage studies also casts further light on the general formulation for the analysis of case-control data outlined in Chapter 4.

Later in this chapter we consider briefly some special sampling designs for matched case-control studies, and in the final part of the chapter we turn to the specific application of family-based case-control studies. These special designs have been developed in particular to address some issues arising in studies of associations between genotypes and disease and also in studies of interactions between genes and environment or lifestyle in explaining disease risk.

6.2 Two-stage sampling design

6.2.1 Preliminaries

There is a variety of two-or-more stage designs in which at least one stage, but not necessarily all stages, are of case-control form. Broadly, more detailed information is collected at the later stages. Individuals sampled at each later stage are a subset of those studied at the previous stage.

We consider two main forms of two-stage sampling designs involving a case-control sample at one or both stages.

(1) In the first form of sampling design, the first-stage sample is a random sample of individuals from an underlying population of interest. Within the first-stage sample the outcome of interest, denoted Y, is observed for each individual, therefore providing information on the proportions of cases and controls in the underlying population. The second-stage sample is a case-control sample from the first stage. Full exposure and covariate information is collected within the case-control sample. As an extension, other information may be available on individuals in the first-stage sample, such that these individuals can be divided into strata, and then the second stage is a stratified case-control sample.

(2) In the second form, both stages of sampling are case-control samples. At the first stage a case-control sample is taken and some, but not all, information on exposures or adjustment variables is obtained. The second stage is a stratified case-control sample, within which some strata defined by outcomes, exposures or adjustment variables are over-represented. Full exposure and covariate information is collected within the second-stage case-control sample.

As noted earlier, a major aim of two-stage sampling is to gain efficiency, that is, to obtain more precise estimates of the primary relationships of interest. Suppose that we wish to estimate the association between a quite rare exposure and an outcome, adjusted for confounders. Under a case-control study with random sampling of cases and controls from the underlying subpopulations, low precision may result. Efficiency may be gained by over-representing exposed individuals. This can be achieved in a two-stage study if some exposure information is available in the first-stage sample, as in the stratified version of the first form of design described above and also as in the second form. Furthermore, two-stage studies can be useful when an interaction between two exposures is of primary interest and where perhaps one of these exposures occurs quite rarely in the underlying population.

6.2.2 A general simplification

A general and relatively simple approach to the analysis of two-stage studies is possible, as follows. The full likelihood can be calculated as a product of the first-stage likelihood and the conditional likelihood of the second stage given the first. Thus not only can the two stages be analysed separately but also the consequent estimates are asymptotically independent. If necessary the combining of contributions from the two stages can be achieved, for instance by weighted least squares, and mutual consistency of the parts thereby checked.

The first-stage analysis is arranged to estimate a parameter $\theta^{(1)}$ of interest by a method appropriate for that stage, leading to an estimate $\hat{\theta}^{(1)}$ with covariance matrix Ω_{11}. The estimation is based on the first-stage likelihood. For the second stage, it is assumed that full information from the first stage is regarded as given, and so the second-stage likelihood may be formed conditionally on the first-stage information. The combined likelihood is therefore the product of two factors. Suppose that the second-stage likelihood is determined by $\theta^{(2)}$, a distinct parameter with a different interpretation from that specifying the first stage, namely $\theta^{(1)}$. The simplest situation is where $\theta^{(1)}$ and $\theta^{(2)}$ are variation independent, that is, any value of one may occur with any value of the other. Then, because of the form of the joint likelihood, the sampling errors in $\hat{\theta}^{(1)}$ and $\hat{\theta}^{(2)}$ are asymptotically independent with covariance matrix, say

$$\text{diag}(\Omega_{11}, \Omega_{22}). \qquad (6.1)$$

It follows that if a parameter vector of interest, $\phi = \phi(\theta^{(1)}, \theta^{(2)})$, is formed from the two separate stage components then the covariance matrix of $\hat{\phi}$ is

$$(\nabla_1^T \phi)\Omega_{11}(\nabla_1 \phi) + (\nabla_2^T \phi)\Omega_{22}(\nabla_2 \phi). \tag{6.2}$$

Here the gradients are calculated with respect to θ_1 and θ_2 respectively.

A more complicated possibility occurs when one cannot parameterize in such a way that the two parameter vectors are variation independent. For example, the formulation may be written in terms of three variation-independent components, namely

$$\theta^{(1)} = (\theta^{(11)}, \theta^{(12)}), \quad \theta^{(2)} = (\theta^{(22)}, \theta^{(21)}), \tag{6.3}$$

where the common part of the covariance matrix is assumed to be such that $\theta^{(12)} = \theta^{(21)}$; consistency with this can and should be used as a partial check of model adequacy. A point estimate is now found by regarding $(\hat{\theta}^{(11)}, \hat{\theta}^{(12)}, \hat{\theta}^{(22)}, \hat{\theta}^{(21)})^T$ as an observation vector with expectation $(\theta^{(11)}, \theta^{(12)}, \theta^{(22)}, \theta^{(12)})^T$ and with a covariance matrix that can be calculated.

In a later section we use this general approach for the two forms of two-stage study described above.

6.2.3 The simplest situation

A simple supplementation of a case-control study is made by adding an independent random sample from the underlying population, estimating solely the population proportion of cases. An alternative is to estimate this proportion from an initial sample and then to form, as a second stage, a case-control study by choosing n_0 controls and n_1 cases as a random sample from within the first stage. The full likelihood is then the product of the likelihood from the first stage and the conditional likelihood of the second stage given the first. The two stages may be analysed separately and, for example to estimate an absolute risk, the estimated population proportion of cases from the first stage may then be combined with the odds ratio estimated from the case-control study. Estimates from the two factors of the likelihood have, as noted above, asymptotically independent sampling errors. The estimation of absolute risks using case-control data supplemented with a random sample from the underlying population was discussed in Section 4.7. This simple type of two-stage design arises when there is a pre-existing cohort (stage 1) that may be under study for a range of different outcomes, within which a case-control study can be taken (stage 2).

If the stage-1 data were found to contain additional information overlapping with that from the case-control sample then in principle, a full likelihood formulation for the whole data would be required to make the best use of the available data. This is outlined in Section 6.3.

6.2.4 Stratified sampling at stage 2

An important extension to the simple design described above is possible when information is available in the first-stage sample about an exposure or adjustment variable of interest dividing the population into strata, thus allowing the use of stratified sampling at the second stage. The two stages are specified as follows.

(1) The stage-1 sample is again a random sample from an underlying population of interest. The outcome variable is observed for all individuals in the sample, N_1 and N_0 denoting the total numbers of cases and non-cases observed. Additionally, certain variables are observed for all individuals in this sample in order to divide the sample into S strata, say. The numbers of cases and controls within stratum s are denoted $N_{1;s}$ and $N_{0;s}$ respectively ($s = 1, \ldots, S$).

(2) At stage 2, $n_{1;s}$ cases and $n_{0;s}$ controls are sampled within each stratum $s = 1, \ldots, S$. All exposures and additional covariates of interest are obtained for all individuals in this stratified case-control sample.

For a single categorical exposure X observed in the stage-2 sample, we let $r_{xy;s}$ denote the number of individuals in stratum s in the case-control sample with exposure $X = x$ ($x = 0, \ldots, J$) and outcome $Y = y$ ($y = 0, 1$). The data can be displayed as in Table 6.1.

Perhaps the more important use of stratified two-stage sampling is that, by adjusting the sampling fractions within different strata, it enables improved study of exposure-outcome associations within rare strata, from which few individuals would be selected to a case-control study under random sampling. It also enables the estimation of the proportions of cases and controls both in the underlying population and also within strata.

There are broadly two approaches to the analysis of such stratified data. The first is useful when there are a small number of strata, each large enough to have a viable analysis and interpretation on its own. In particular the strata may be domains of study. That is, separate conclusions for individual domains or groups of domains may be of interest. The combining of estimates across strata was discussed in Section 2.8. The second approach is used when there are many strata, each individually relatively uninformative. Stratified analyses using the generalized formulation were outlined in

Table 6.1 *Data from a two-stage sample with random sampling at stage* 1 *and stratified case-control sampling at stage* 2.

(a) Stage 1: Random sample.

	\multicolumn{4}{c}{Stratum}			
	1	2	$\cdots$	S	Total
$Y = 0$	$N_{0;1}$	$N_{0;2}$	$\cdots$	$N_{0;S}$	N_0
$Y = 1$	$N_{1;1}$	$N_{1;2}$	$\cdots$	$N_{1;S}$	N_1
Total	$N_{.;1}$	$N_{.;2}$		$N_{.;S}$	

(b) Stage 2: Stratified case-control sample.

	\multicolumn{3}{c}{Stratum 1}	$\cdots$	\multicolumn{3}{c}{Stratum S}
	$X = 0$	$\cdots$ $X = J$	Total	$\cdots$	$X = 0$	$\cdots$ $X = J$	Total
$Y = 0$	$r_{00;1}$	$\cdots$ $r_{J0;1}$	$n_{0;1}$		$r_{00;S}$	$\cdots$ $r_{J0;S}$	$n_{0;S}$
$Y = 1$	$r_{01;1}$	$\cdots$ $r_{J1;1}$	$n_{1;1}$		$r_{01;S}$	$\cdots$ $r_{J1;S}$	$n_{1;S}$
Total	$r_{0.;1}$	$r_{J.;1}$	$n_{.;1}$		$r_{0.;S}$	$r_{J.;S}$	$n_{.;S}$

Section 4.3. Likelihood analyses for two-stage studies will be described in Section 6.3.

6.2.5 Two stages with case-control samples at both

Here we extend the two-stage sampling design to a situation where the first-stage sample is a case-control rather than a random sample. The second stage is thus a further case-control sample within the first stage, in which some groups of individuals are over-represented.

At the first stage, information on certain exposures is obtained in the case-control sample. In the second stage fixed numbers of individuals are sampled within strata defined by a cross classification of case-control status with a further categorical variable or by a categorized continuous variable, which is observed in the first-stage sample. Full exposure and covariate information is then obtained for all individuals in the second-stage sample.

We suppose that our interest is in the association of the outcome Y with a main exposure, X, controlling for an additional variable W. However, the roles of X and W could be reversed. Suppose that X is observed in the first-stage case-control sample. If some values of X are relatively

Table 6.2 *Data from a two-stage case-control sample with a standard case-control sample at stage 1 and stratified case-control sample at stage 2, for a binary main exposure X and single categorical covariate W.*

(a) Stage 1.

	$X = 0$	$X = 1$	Total
$Y = 0$	N_{00}	N_{10}	N_0
$Y = 1$	N_{01}	N_{11}	N_1

(a) Stage 2.

	$W = 1$		$\cdots$	$W = K$		Total	
	$X = 0$	$X = 1$	$\cdots$	$X = 0$	$X = 1$	$X = 0$	$X = 1$
$Y = 0$	$r_{00;1}$	$r_{10;1}$	$\cdots$	$r_{00;K}$	$r_{10;K}$	n_{00}	n_{10}
$Y = 1$	$r_{01;1}$	$r_{11;1}$	$\cdots$	$r_{01;K}$	$r_{11;K}$	n_{01}	n_{11}

infrequently observed in the first-stage sample, because they occur rarely in the underlying population, then individuals with these values may be over-represented in the second-stage sample. Variable W is then observed in the second stage sample only.

The data from this type of two-stage study are shown in Table 6.2.

Example 6.1: Two-stage study with case-control sampling at both stages
Hanley *et al.* (2005) outlined a two-stage case-control study of the association between vasectomy and myocardial infarction. The first stage involved a sample of 143 men who had suffered a myocardial infarction, the cases, and 1430 controls. Of the cases, 23 had had a vasectomy; the corresponding number among the controls was 238. The log odds ratio for the association was log 0.96 with estimated standard error 0.0239 (95% confidence interval -0.088 to 0.006). There was a concern that smoking status could have confounded the association but it was too expensive to obtain detailed smoking histories for all 1573 men in the case-control sample. A second-stage sample was therefore taken, and for this the numbers in each of the four vasectomy–myocardial-infarction groups were chosen to be about equal: in fact 20 cases were selected with vasectomy and 16 without, and 16 controls were selected with vasectomy and 20 without. Smoking status was ascertained in the second-stage sample and the estimated smoking-adjusted log odds ratio for the association between vasectomy and myocardial infarction was log 1.09 with estimated standard error 0.028 (95% confidence interval

0.031 to 0.141). Before the adjustment for smoking, an inverse, though non-statistically-significant, association was found between vasectomy and myocardial infarction. After adjusting for smoking status, vasectomy was associated with increased risk of myocardial infarction and the association was, although relatively small, statistically significant at the 0.05 level. If random samples of cases and controls had been sampled at the second stage, this could have resulted in too few men in the second-stage sample who had had a vasectomy and a statistically significant adjusted odds ratio estimate might not have been found.

Special analyses are required for two-stage designs with case-control sampling at both stages. As noted above, suppose we are primarily interested in the association between X and Y, adjusted for W, where W is a categorical variable. The quantity of interest may be taken to be β, the conditional odds ratio given $W = w$ in the model

$$\Pr(Y = 1 | X = x, W = w) = L_1(\alpha + \beta x + \gamma_w). \tag{6.4}$$

Here β specifies the dependence on x adjusted to give the odds ratio at any fixed value of w, and the probabilities refer to the underlying population.

For a binary exposure X, the odds ratio within the group with $W = w$ ($w = 1, \ldots, K$) is

$$\zeta_w = \frac{\Pr(Y = 1 | X = 1, W = w)/\Pr(Y = 0 | X = 1, W = w)}{\Pr(Y = 1 | X = 0, W = w)/\Pr(Y = 0 | X = 0, W = w)}. \tag{6.5}$$

By Bayes' theorem this can be written

$$\zeta_w = \frac{\Pr(X = 1 | Y = 1, W = w)/\Pr(X = 0 | Y = 1, W = w)}{\Pr(X = 1 | Y = 0, W = w)/\Pr(X = 0 | Y = 0, W = w)}. \tag{6.6}$$

The probabilities in the above odds ratios refer to the underlying population of interest. Under standard case-control sampling the probabilities $\Pr(X = 1 | Y = 1, W = w)$ can be estimated directly from the case-control data, as discussed in Chapters 2 and 4. If, under the two-stage design, both X and W were observed in the stage-1 case-control sample then the probabilities could be estimated directly within that sample data. However, under the two-stage design these probabilities cannot be estimated directly using only the second-stage data.

Instead, we can rearrange the odds ratio in (6.5) in the form

$$\zeta_w = \frac{\Pr(W = w | Y = 1, X = 1)\Pr(W = w | Y = 0, X = 0)}{\Pr(W = w | Y = 1, X = 0)\Pr(W = w | Y = 0, X = 1)}$$
$$\times \frac{\Pr(Y = 1 | X = 1)/\Pr(Y = 0 | X = 1)}{\Pr(Y = 1 | X = 0)/\Pr(Y = 0 | X = 0)}$$
$$= \phi_w \zeta, \tag{6.7}$$

say, where ζ is the marginal, or unadjusted, odds ratio and ϕ_w is a correction factor. The probabilities $\Pr(W = w | Y = y, X = x)$ can be estimated within the stage-2 case-control data, and the odds ratio that is the second factor can be estimated within the stage-1 data.

For the data in Table 6.2 the group-specific odds ratios are therefore estimated by

$$\hat{\zeta}_w = \hat{\phi}_w \hat{\zeta} = \frac{(r_{11;w}/n_{11})(r_{00;w}/n_{00})}{(r_{01;w}/n_{01})(r_{10;w}/n_{10})} \frac{N_{11} N_{00}}{N_{01} N_{10}}. \tag{6.8}$$

The adjusted log odds ratio $\beta = \log \zeta_{\mathrm{adj}}$, say, has generalized least squares estimate

$$\hat{\beta} = (A^T V^{-1} A)^{-1} A^T V^{-1} \hat{B}, \tag{6.9}$$

where A is a $K \times 1$ column vector of 1s, $\hat{B} = (\hat{\beta}_1, \ldots, \hat{\beta}_K)^T$ with $\hat{\beta}_w = \log \hat{\zeta}_w$ and V is the $K \times K$ variance-covariance matrix for the estimated stratum-specific log odds ratios $\hat{\beta}_w$ $(w = 1, \ldots, K)$. The variance of the estimate $\hat{\beta}$ is $(A^T V^{-1} A)^{-1}$. The elements of V are

$$v_w = \mathrm{var}(\hat{\beta}_w) = \mathrm{var}\{\log(\hat{\phi}_w)\} + \mathrm{var}\{\log(\hat{\zeta})\} \tag{6.10}$$

$$c_{ww'} = \mathrm{cov}\{\log(\hat{\zeta}_w), \log(\hat{\zeta}_{w'})\}. \tag{6.11}$$

From previous results we have

$$\mathrm{var}\{\log(\hat{\zeta})\} = \frac{1}{N_{11}} + \frac{1}{N_{00}} + \frac{1}{N_{01}} + \frac{1}{N_{10}}. \tag{6.12}$$

To find the variance of $\log \hat{\phi}_w$, first let

$$\log \hat{\phi}_w = \log \frac{r_{11;w}}{n_{11}} + \log \frac{r_{00;w}}{n_{00}} + \log \frac{r_{10;w}}{n_{10}} + \log \frac{r_{01;w}}{n_{01}}$$

$$= g(r_{11;w}, r_{00;w}, r_{01;w}, r_{10;w}). \tag{6.13}$$

We now use the result that for large samples the variance-covariance matrix for $g(r_{11;w}, r_{00;w}, r_{01;w}, r_{10;w})$ is $G \Sigma G^T$, where G is the vector of derivatives of $g(\cdot)$ with respect to $\{r_{11;w}, r_{00;w}, r_{01;w}, r_{10;w}\}$ and Σ is the estimated variance-covariance matrix for $\{r_{11;w}, r_{00;w}, r_{01;w}, r_{10;w}\}$, which has off-diagonal elements 0. Note also that for given x and y the $\{r_{xy;1},...,r_{xy;K}\}$ have a multinomial distribution. It follows that

$$v_w = \sum_{x,y} \frac{1}{r_{xy;w}} - \sum_{x,y} \frac{1}{n_{xy}} + \sum_{x,y} \frac{1}{N_{xy}}, \tag{6.14}$$

$$c_{ww'} = \sum_{x,y} \frac{1}{N_{xy}} - \sum_{x,y} \frac{1}{n_{xy}}, \quad w \neq w'. \tag{6.15}$$

Likelihood-based analyses are considered in Section 6.3.3. A large number of strata showing some random variation of ζ between strata may lead to shrinkage of the estimated stratum effects towards their mean.

Example 6.2: A numerical example It may be an advantage to consider a two-stage case-control study with case-control sampling at both stages, instead of a standard one-stage case-control study of similar cost. A detailed comparison must depend on the costs of obtaining exposure and covariate information. One possibility is that the stage-1 information is readily available and hence inexpensive and that it is the stage-2 information which is expensive to obtain. With this in mind we use a numerical example to compare the two-stage study described in this section with a corresponding one-stage study: the numbers of cases and controls are the same in the one-stage study and the second stage of the two-stage study.

The data are shown in Table 6.3. Table 6.3(a) shows full data on a binary outcome Y, a binary exposure X and a binary confounder W from the stage-1 case-control sample; in reality the information on W would be unobserved in the two-stage setting. Table 6.3(b) shows the data obtained from both stages of a two-stage sample: the first stage is as in Table 6.3(a) and the second stage uses balanced sampling in the four (X, Y) groups. Table 6.3(c) shows the data from a comparable one-stage case-control sample.

If we were able to observe both W and X in the first-stage sample in Table 6.3(a) then the odds ratio estimates within the strata defined by W and the pooled estimate would be given by

$$\hat{\beta}_0 = 0.57(0.23), \quad \hat{\beta}_1 = 0.54(0.18), \quad \hat{\beta} = 0.55(0.14),$$

where the standard errors are given in brackets. Under the two-stage sampling design in Table 6.3(b) the estimates and their standard errors are

$$\hat{\beta}_0 = 0.57(0.31), \quad \hat{\beta}_1 = 0.54(0.23), \quad \hat{\beta} = 0.55(0.15)$$

while under the simple case-control study in Table 6.3(c), the estimates are

$$\hat{\beta}_0 = 0.59(0.50), \quad \hat{\beta}_1 = 0.53(0.40), \quad \hat{\beta} = 0.55(0.31).$$

In this example, therefore, for estimating the adjusted log odds ratio β the two-stage case-control study is highly efficient relative to the simple case-control study.

Table 6.3 *Numerical example: Comparison of a two-stage case-control study with a simple case-control sample, for a binary outcome Y, binary exposure X and binary confounder W.*

(a) Stage-1 case-control sample. in reality W is unobserved in this sample, but we show the data for the purposes of comparison.

	$X = 0$			$X = 1$			Total
	$W = 0$	$W = 1$	Total	$W = 0$	$W = 1$	Total	
$Y = 0$	550	350	900	50	50	100	1000
$Y = 1$	250	570	820	40	140	180	1000

(b) A two-stage case-control study where the stage-1 sample is as in (a) but with only X and Y observed. Balanced sampling is used to obtain the second-stage data, using 100 in each of the four (X, Y) groups.

(b1) Stage 1.

	$X = 0$	$X = 1$	Total
$Y = 0$	900	100	1000
$Y = 1$	820	180	1000

(b2) Stage 2.

	$X = 0$			$X = 1$		
	$W = 0$	$W = 1$	Total	$W = 0$	$W = 1$	Total
$Y = 0$	61	39	100	50	50	100
$Y = 1$	30	70	100	22	78	100

(c) A simple case-control sample within the stage-1 sample, with 200 cases and 200 controls.

	$X = 0$		$X = 1$		Total
	$W = 0$	$W = 1$	$W = 0$	$W = 1$	
$Y = 0$	110	70	10	10	200
$Y = 1$	49	115	8	28	200

6.3 Likelihood analyses for two-stage studies

6.3.1 Preliminaries

In this section we discuss the likelihood-based analysis of the full data arising from two-stage sampling designs. It will be shown that two-stage case-control studies of both forms given at the start of Section 6.2.1 can be analysed using modified versions of the general formulation in Chapter 4. The modifications are in the form of 'offset' terms in the logistic regression. What follows from these results is that two-stage case-control studies can be analysed, using logistic regression, as though the data had arisen in a one-stage random prospective sample, with a straightforward adjustment. Logistic regression with offset terms is allowed in commonly used statistical software packages.

6.3.2 Random sample at stage 1

To apply the general procedure outlined in Section 6.2.2 when the first stage involves a random sample from the population and the second stage is of case-control form, possibly stratified, it is relatively straightforward to combine the information across stages and where necessary across strata. We outline this below for the general situation of stratified sampling at stage 2.

We focus on the following situation: in the stage-1 sample information the outcome Y and stratum S are observed for each individual, and at stage 2 individuals are sampled within groups defined by a cross-classification of Y and S and a categorical exposure X is observed. The data are as displayed in Table 6.1. The full likelihood for the two-stage data is the product of factors from the first stage and factors from the second stage conditional on the first stage:

$$\text{lik} = \prod_{y,s} \text{Pr}^{\mathcal{D}_1}(Y = y | S = s)^{N_{y;s}} \prod_{x,y,s} \text{Pr}^{\mathcal{D}_2}(X = x | Y = y, S = s)^{r_{xy;s}}$$

(6.16)

where the superscript $\mathcal{D}_1$ denotes the sampling model for the stage-1 sample and $\mathcal{D}_2$ denotes the sampling model for the stage-2 sample. The likelihood can be rearranged in the form

$$\text{lik} \propto \prod_{y,s} \text{Pr}^{\mathcal{D}_1}(Y = y | S = s)^{N_{y;s}} \prod_{x,y,s} \text{Pr}^{\mathcal{D}_2}(Y = y | X = x, S = s)^{r_{xy;s}}$$

$$\times \prod_{x,y,s} \text{Pr}^{\mathcal{D}_2}(X = x | S = s)^{r_{xy;s}}.$$

(6.17)

Stratification can be incorporated into the logistic regression model either by including a separate intercept term α_s for each stratum or by modelling the stratum effects. We begin by focusing on the first situation, in which the inverse model of interest is

$$\mathrm{Pr}^{\mathcal{I}}(Y = y | X = x, S = s) = L_y(\alpha_s + \beta_x). \tag{6.18}$$

Under this model the term $\mathrm{Pr}^{\mathcal{D}_2}(Y = y | X = x, S = s)$ can be written as

$$L_y\left(\alpha_s + \log \frac{n_{1;s}}{n_{0;s}} - \log \frac{\omega_{1s}}{\omega_{0s}} + \beta_x\right), \tag{6.19}$$

where $\omega_{ys} = \mathrm{Pr}^{\mathcal{D}_1}(Y = 1 | S = s)$.

At stage 2 the numbers of cases and controls sampled within strata, $(n_{1;s}, n_{0;s})$, are regarded as fixed conditional on being in the sample, placing a constraint on $\mathrm{Pr}^{\mathcal{D}_2}(Y = y)$, which can be expressed as the constraint that, for $y = 0, 1$,

$$n_{y;s} - n \sum_x \mathrm{Pr}^{\mathcal{D}_2}(Y = y | X = x, S = s) \mathrm{Pr}^{\mathcal{D}_2}(X = x | S = s) = 0. \tag{6.20}$$

The likelihood (6.17) can be maximized subject to the constraints by introducing Lagrange multipliers from (6.20).

It can be shown that the score equations for (α_s, β_x) using the full constrained likelihood in (6.17) are the same as those that would be obtained from an analysis based only on the second term of the likelihood, with the probabilities ω_{ys} replaced by their estimates from the stage-1 sample, $N_{y;s}/N_{.;s}$; that is, (α_s, β_x) can be estimated using the *pseudo-likelihood*

$$\prod_{x,y,s} L_y\left(\alpha_s + \log \frac{n_{1;s}}{n_{0;s}} - \log \frac{N_{1;s}}{N_{0;s}} + \beta_x\right)^{r_{xy;s}}. \tag{6.21}$$

Therefore the parameters (α_s, β_x) can be estimated by a logistic regression of the stage-2 case-control data, using the following fixed offset within each stratum in the regression:

$$\log \frac{n_{1;s}}{n_{0;s}} - \log \frac{N_{1;s}}{N_{0;s}}. \tag{6.22}$$

Pseudo-likelihoods are discussed in the appendix. The variance of the resulting estimates $(\hat{\alpha}_s, \hat{\beta}_x)$ is found using the sandwich formula and is given by

$$\hat{I}^{-1} - \begin{pmatrix} D & 0 \\ 0 & 0 \end{pmatrix} \tag{6.23}$$

where $\hat{I}$ is the information matrix from the case-control logistic regression using the offsets given by (6.22) and D is a matrix with diagonal entries $1/n_{1;s} + 1/n_{0;s} - 1/N_{1;s} - 1/N_{0;s}$ and zeros elsewhere.

The above link between the full likelihood for the two-stage data and the part of the likelihood referring only to the stage-2 case-control data is another way of arriving at the general formulation results of Chapter 4.

The situation is different when the stratum effects are modelled using a regression term. In this case the inverse model of interest is

$$\Pr^{\mathcal{I}}(Y = y | X = x, S = s) = L_y(\alpha + \beta_x + \gamma z_s), \qquad (6.24)$$

where the z_s denote variables that define stratum s. Here the relationship between the part of the likelihood based on the second-stage case-control data only and the full likelihood in (6.17) is more complicated. As above, Lagrange multipliers can be introduced to maximize the full likelihood with respect to the constraints.

The score equations for α, β_x, γ are the same as those that would arise from a likelihood analysis based on an analysis using only the stage-2 case-control data, except that the intercept α is replaced in stratum s by $\alpha + \xi_s$ where

$$\xi_s = \log \frac{n_{1;s} - T_s}{n_{0;s} + T_s} + \log \frac{\omega_{1s}}{\omega_{0s}} \qquad (6.25)$$

and $T_s = N_{1;s} - N_{.;s} \Pr^{\mathcal{D}_1}(Y = 0 | S = s)$. Here, the equation for ξ_s contains the parameter ω_{ys}, for which in this situation there is no closed form estimate. As a result the score equations must be solved iteratively. Choosing the value $\omega_{ys} = N_{y;s}/N_{.;s}$ gives estimates of $(\alpha, \beta_x, \gamma)$ that are the same as would be obtained by fitting a logistic regression to the second-stage case-control data using a fixed offset in each stratum; here the offset is ξ_s with ω_{ys} replaced by the stage-1 estimate $N_{y;s}/N_{.;s}$. However, these estimates are not in general maximum likelihood. A correction to the information matrix from an analysis based only on the case-control data is also required to obtain correct variance estimates for $(\alpha, \beta_x, \gamma)$. The details of this are not given here.

In summary, when stratum effects are modelled the odds ratio estimates that would be obtained under a pseudo-likelihood analysis of the stage-2 case-control data are *not* the same as those that would arise under a full likelihood analysis of the combined two-stage data. However, these estimates can be obtained by iteratively fitting the pseudo-likelihood product that constitutes the second factor in (6.17), which is a logistic regression with an offset.

No matter how stratification is incorporated into the logistic regression model, it is possible that there is something to be gained in terms of the precision of parameter estimates by an analysis of the complete two-stage data using the full likelihood in (6.17) rather than the pseudo-likelihood analyses described here. However, such an analysis is considerably more complex and may not be generally applied in the available statistical software. Special variance estimation is required to deal with the constraints under a full-likelihood analysis. We will not discuss full-likelihood analyses further here; we refer to the notes for authors who have investigated this.

6.3.3 Case-control studies at both stages

The likelihood results outlined above can be extended to the more complex setting of a two-stage study with case-control sampling at both stages. A 'simple' analysis for data arising under this design was given in (6.25). However, those methods become cumbersome when a number of covariates W are obtained at the second stage of a two-stage case-control study or if W is continuous. Here we outline a more general likelihood-based analysis.

To simplify the discussion we assume a binary exposure X and a categorical covariate W, with the data arranged as in Table 6.2. The full likelihood for the observed data is again the product of the likelihoods from the two stages:

$$
\text{lik} \propto \prod_{x,y} \Pr^{\mathcal{D}_1}(X = x | Y = y)^{N_{xy}}
$$
$$
\times \prod_{x,y,w} \Pr^{\mathcal{D}_2}(W = w | X = x, Y = y)^{r_{xy;w}}, \qquad (6.26)
$$

where $\mathcal{D}_1$ denotes the sampling model for an individual's being in the stage-1 sample and $\mathcal{D}_2$ denotes the corresponding model for the stage-2 sample. To proceed it is helpful to rearrange the likelihood in the form

$$
\text{lik} \propto \prod_{x,y} \Pr^{\mathcal{D}_1}(Y = y | X = x)^{N_{xy}} \prod_{x} \Pr^{\mathcal{D}_1}(X = x)^{N_{x.}}
$$
$$
\times \prod_{x,y,w} \Pr^{\mathcal{D}_2}(Y = y | X = x, W = w)^{r_{xy;w}}
$$
$$
\times \prod_{x,w} \Pr^{\mathcal{D}_2}(X = x, W = w)^{r_{x.;w}}. \qquad (6.27)
$$

The parameters of the likelihood can be estimated subject to the constraints that, on condition that an individual is in the stage-1 or stage-2 samples, N_y and n_{xy} are fixed. These constraints can be expressed, for $y = 0, 1$ and $x = 0, \ldots, J$, as

$$N_y - N \sum_x \text{Pr}^{\mathcal{D}_1}(Y = y | X = x)\,\text{Pr}^{\mathcal{D}_1}(X = x) = 0, \quad (6.28)$$

$$n_{xy} - n \sum_w \text{Pr}^{\mathcal{D}_2}(Y = y | X = x, W = w)$$

$$\times \text{Pr}^{\mathcal{D}_2}(X = x, W = w) = 0. \quad (6.29)$$

We assume, for the association of interest in the inverse model,

$$\text{Pr}^{\mathcal{I}}(Y = y | X = x, W = w) = L_y(\alpha + \beta_x + \gamma_w). \quad (6.30)$$

Under this model it can be shown that

$$\text{Pr}^{\mathcal{D}_1}(Y = y | X = x) = L_y\{v_x + \log(N_1/N_0)\}, \quad (6.31)$$

$$\text{Pr}^{\mathcal{D}_2}(Y = y | X = x, W = w)$$

$$= L_y\{\log(n_{x1}/n_{x0}) + \alpha^* + \beta_x + \gamma_w - v_x\}, \quad (6.32)$$

where

$$v_x = \log\{\text{Pr}^{\mathcal{I}}(X = x | Y = 1)/\text{Pr}^{\mathcal{I}}(X = x | Y = 0)\} \quad (6.33)$$

and

$$\alpha^* = \alpha + \log\{\text{Pr}^{\mathcal{I}}(Y = 0)/\text{Pr}^{\mathcal{I}}(Y = 1)\}. \quad (6.34)$$

There is no information in the two-stage data described here about the marginal probability of being a case for an individual in the underlying population. This is why the probabilities $\text{Pr}^{\mathcal{I}}(Y = y)$ have been absorbed into the parameter α^*.

As in the previous section, the likelihood (6.27) can be maximized subject to the constraints in (6.28) and (6.29) by introducing a set of Lagrange multipliers. We omit the details, but it can be shown that the maximum likelihood estimates for $\text{Pr}^{\mathcal{D}_1}(X = x)$, $\text{Pr}^{\mathcal{D}_2}(X = x, W = w)$ and v_x are the unconstrained estimates

$$\widehat{\text{Pr}}^{\mathcal{D}_1}(X = x) = N_{x.}/N, \quad \widehat{\text{Pr}}^{\mathcal{D}_2}(X = x, W = w) = r_{x.;w}/n,$$

$$\hat{v}_x = \log\{(N_{x1}N_0)/(N_{x0}N_1)\}.$$

The score equations arising from the full likelihood (6.27) can therefore be shown to be the same as those that would arise from an unconstrained maximization of just the third term of the likelihood, which is a pseudo-likelihood given by

$$\prod_{x,y,w} L_y \{\log(n_{x1}/n_{x0}) + \alpha^* + \beta_x + \gamma_w - \hat{v}_x\}^{r_{xy;w}}. \tag{6.35}$$

The parameters $(\alpha^*, \beta_x, \gamma_w)$ can thus be estimated using a logistic regression based on the stage-2 data with an offset term

$$\log(n_{x1}/n_{x0}) - \log\{(N_{x1}N_0)/(N_{x0}N_1)\}. \tag{6.36}$$

The information matrix does not give the covariance matrix of the estimates. The more elaborate calculation for this is omitted; it gives for a binary exposure X observed at stage 1 the variance of the parameter of interest as

$$\text{vâr}(\hat{\beta}) = I_{\beta\beta}^{-1} - \sum_{x,y} 1/n_{xy} + \sum_{x,y} 1/N_{xy}, \quad x, y = 0, 1, \tag{6.37}$$

where $I_{\beta\beta}$ denotes the diagonal element for β in the information matrix from the second-stage pseudo-likelihood (6.35). The variances of the parameters associated with the covariates observed at stage 2 are correctly estimated by the appropriate term in the inverse of the information matrix from the pseudo-likelihood. Any terms representing interactions between first- and second-stage variables are also estimated correctly using the pseudo-likelihood, no correction being required to obtain the correct variances.

The above results refer specifically to the situation in which a separate parameter γ_w is estimated for each level of the factor observed in the second-stage sample. However, when the effects of the covariates W are modelled in some way in $\Pr(Y = y | X = x, W = w)$, the situation is more complicated. We omit the details here, noting that they are similar to those given in Section 6.3.2. In this situation, therefore, the estimates obtained from a pseudo-likelihood analysis in which we maximize the third product of the full likelihood (6.27) are *not* maximum likelihood estimates. However, the corresponding maximum likelihood estimates can be obtained by the iterative fitting of a logistic regression equivalent to the third product in the full likelihood (6.27).

6.4 Discussion

6.4.1 Use of two-stage studies

Large assembled cohorts or large databases of individuals for which some limited information is available for all individuals provide ideal settings within which two-stage case-control designs with stratified sampling at the second stage can be used. In particular, these designs have important potential applications in studies of expensive biological exposures, for example genetic exposures, where the biological material may have been obtained and stored for all individuals in a 'biobank'. In such studies the biological material is a finite resource (that is, it cannot be processed repeatedly) and the processing to obtain measurements of interest may be expensive.

Case-control sampling designs that make use of information available in the pool of potential controls on certain exposures have an important application in the study of gene-environment interactions, where the environmental exposure, or perhaps a surrogate of it, is cheaply available.

6.4.2 *Optimal two-stage sampling*

In two-stage studies involving the stratified sampling of individuals at the second stage, the question arises how to sample individuals optimally within strata. We focus on the two-stage design with case-control sampling at both stages, with (Y, X) observed at stage 1 and W additionally observed at stage 2. The simplest possibility is that a *balanced sampling design* is used, in which the same number of individuals is sampled in each (Y, X) subgroup to form the second-stage sample. This was the procedure used in Example 6.2. It may not always result in a gain in efficiency relative to a standard one-stage case-control study. However, there may exist a better way of choosing the numbers to sample at the second stage. If the main interest lies in the adjusted association between X and Y given W, measured using an odds ratio, then we should choose the n_{xy} to minimize the variance of the pooled log odds ratio estimate. To approximate these optimal numbers requires some prior knowledge of the expected associations among (X, Y, W). In many circumstances such information is not likely to be available, however, and then a balanced design may be used as a compromise.

A balanced two-stage sample may be inefficient even relative to standard sampling if W is not a confounder but any loss in efficiency is likely to be small. A balanced design has also been found in most circumstances to have an efficiency close to unity relative to the optimal design, for both the main term of interest and for interactions. The estimated variances of odds ratio estimates for the main effects of variables observed only at stage 2 may be larger than they would have been under standard sampling.

6.4.3 *Multistage studies*

Two-stage case-control designs can be extended in a natural way to studies with multiple stages, for example to a three-stage design that in effect combines the two forms of two-stage design considered so far. In this three-stage design the first stage is a random sample from the underlying population, providing estimates of the proportions of cases and controls. The second stage is a standard case-control sample within which information on an important exposure or confounder is obtained, and the third stage is a stratified case-control sample within which full exposure and covariate information is obtained. Likelihood analyses arise as extensions to the above workings.

Example 6.3: A case-control study with multi-stage sampling Benichou *et al.* (1997) used a three-stage sample in a study of risk factors for breast cancer. The aim of the study was to estimate exposure-specific rates and in particular to improve existing estimates by including, for example, additional information on mammographic density. In this example the stage-1 sample was an established large cohort of women recruited when free of breast cancer, who then attended breast cancer screening annually for five years. The second stage was a matched case-control sample in which information on several risk factors of interest was obtained. Mammographic density based on records from the first breast cancer screening of the underlying study was to be evaluated in a third-stage sample. The second-stage sample was restricted to a subset of participating study centres and to cases occurring later than the first year of follow up plus their matched controls. This is an example where the sampling fractions at the second stage were determined by practical constraints rather than being set by the investigators. The authors used a Poisson model for the number of cases in the full cohort and a pseudo-likelihood approach that took into account the sampling design in order to estimate the quantities of interest.

6.5 Special sampling in matched case-control studies

The two-stage sampling designs discussed in the previous sections in this chapter are useful extensions to unmatched case-control studies. Example 6.3 illustrated one way in which this approach can be used in the setting of a matched case-control study. We now consider briefly some alternative designs that allow us to make a more efficient use of resources in matched case-control studies.

First we note that if a main exposure of interest is rare in the underlying population then a matched case-control study sampled within that population may have many matched sets that are concordant, that is, sets in which all individuals have the same level of exposure. Such sets are uninformative under the conditional analysis for matched case-control studies. The problem is reduced as the numbers of controls matched to each case is increased. However, the number of controls per case will usually be restricted by constraints of cost or resources. Even for exposures that are relatively common in the underlying population, matched case-control studies with just one control per case, say, would be expected to have a non-negligible number of concordant pairs.

A simple possible two-stage approach for matched studies is as follows:

(1) At the first stage of sampling a standard matched case-control sample is obtained and an exposure of interest is observed for individuals in the sample. We focus on a binary exposure.
(2) The stage-2 sample is restricted to only those matched sets from stage 1 that are informative about the difference in exposure between cases and controls, that is, those matched sets that are discordant with respect to the binary exposure. Additional information on adjustment variables is obtained for individuals in the stage-2 sample.

This approach is suited to a situation in which the main exposure of interest is fairly inexpensive to measure, but where adjustment variables may be expensive to measure. This was the situation in Example 6.1, where the main exposure was having had a vasectomy and it was of interest to adjust for smoking history. Suppose that the cost of the first-stage sample is small, and so the primary cost is of obtaining the information at stage 2. Then this approach would be expected to yield more efficient estimates than a corresponding one-stage study of similar size and therefore similar cost. The analysis of this type of two-stage matched study is the same at that of any matched study.

As discussed in earlier sections in this chapter, in some case-control studies there is available some basic covariate information for individuals in the population from which the sample is drawn, the *sampling population*. As a slight extension, a random sample from the sampling population may be available, for example in the form of an assembled cohort within which some basic information is readily available. Counter-matching and quota sampling are sampling designs for matched case-control studies that make use of this information.

- *Counter-matching* uses covariate information from the population to influence the sampling of controls for each case in such a way that the variability of the exposure is, as far as is possible using the information available, maximized within each matched set. Broadly, if a case is observed to have a certain explanatory variable then counter-matching is used to select matched controls for that case with different values of that explanatory variable.

- *Quota sampling* is an alternative to counter-matching that can be used when it is not feasible to observe any variables of interest in the sampling population. Under this design controls are sampled sequentially into each matched set until a quota of individuals with each value of X is achieved within each set. The variable X is only ascertained for each control after selection.

Both the counter-matching and quota sampling designs require a modification to the usual conditional analysis. These sampling designs were first considered for use in nested case-control studies, and we discuss both in further detail, including the special form of analysis needed, in Chapter 7.

6.6 Case-control studies using family members

6.6.1 Preliminaries

Finally, we discuss briefly a quite different type of special case-control sampling, the use of family groupings. There are a number of ways in which family groupings have been used in case-control studies and a growing body of literature exists in the epidemiological, statistical and genetics fields. Here we discuss two different ways in which data on family members can be used in case-control studies.

- In the first, non-diseased relatives are used as controls for cases.
- In the second, cases and controls are recruited as in a 'standard' case-control study and then the data are augmented by recruiting one or more family members of each case and each control, and their case or control status and exposures and covariates are also observed.

6.6.2 Family members as matched controls

The case-control designs discussed so far have focused on the use of cases with unrelated individuals as controls. There are some situations, however, when it can be advantageous to recruit non-diseased members from the family of each case to serve as controls.

The primary reason for using family members of cases as controls is that cases and controls will be closely matched on 'genetic background', which is of importance in some studies. The degree of matching on genetic backgrounds depends on which type of relative is selected, an aspect which we expand on below.

Family-matched case-control studies are becoming a popular tool in genetic epidemiology, that is, in studies of relationships between specific genotypes and disease outcomes. They are also used in epidemiological studies of gene-environment interactions. Family-matched studies could also be used in case-control studies in other areas, for example sociology, where it may be of interest to disentangle genetic and environmental factors. In all such studies it is important to control for background genetic features that may be related to both the exposure and the outcome. The control group in a case-control study should be sampled from the same underlying population as that in which the cases arose. If the outcome of interest is thought to have a genetic component that is not of specific interest, then this implies that cases and controls should be chosen to have similar genetic backgrounds. If the sampling results in different distributions of certain genetic features within cases or controls then spurious associations may well arise in studies of specific genotypes. This is referred to as *population stratification* or *genetic admixture*; it amounts essentially to confounding by genetic background. One way to try to avoid this is to match cases to controls on the basis of race or ethnicity. However, while often this is likely to be adequate in a diverse population, it may leave remaining systematic differences between the genetic backgrounds of cases and controls. Using family members as matched controls can therefore help when it is known or suspected that there are strong genetic factors at work.

A practical benefit to using family members of cases as controls is that they may be particularly willing to participate in the study, more so than population controls. However, the availability of relatives is an important consideration. Both the degree of control for genetic factors and the feasibility of a study depend on the precise family relationship between each case and his or her control(s).

In a *case-sibling design*, one or more siblings of a case are chosen to serve as matched controls. This aims to give exact matching on ethnicity, thus ensuring that the case and its controls arise from the same genetic population. A drawback of the design is that not all cases will have siblings and omission of those without siblings could result in substantial loss of information or even bias. Care must also be taken in deciding which siblings should be selected as controls. For example, selection of a younger sibling who might later develop the outcome under study appears inappropriate and relates to the common need to incorporate 'time' into the selection of controls. For this reason, it has been suggested that only older siblings should be sampled as controls. This restricts the study further to those with a living older sibling.

Matching cases to first cousins is an alternative to using siblings. The *case-cousin design* is likely to result in fewer cases being dropped owing to the lack of an available control, and it can also allow for closer matching of the case to the cousin on age. However, in comparison with the case-sibling design, under the case-cousin design cases are not of course as closely matched to their control(s) on genetic background, which could result in residual confounding.

Case-sibling and case-cousin studies can be analysed using conditional logistic regression. What is gained in confidence that confounding by genetic background has been avoided by the use of related controls may be paid for in a loss of efficiency in studies of genotype effects compared with that obtained using population controls. The reason is that using related controls, especially siblings, is likely to result in more concordant and therefore uninformative matched sets. However, family-matched studies can enable more precise estimation of gene-environment effects than the use of population controls, in particular when a genotype of interest is rare in the underlying population.

Studies using spouses and children of cases as controls have also been described. Another alternative is to use information about the genotype of the parents of each case. In a design referred to as a *case-parent* or *case-pseudosib* study, each case is 'matched' to the other possible 'pseudo-siblings' that could have arisen from the combination of the parents' genotypes. If the disease is rare then it may be assumed that the pseudo-siblings would have been controls and the analysis is by conditional logistic regression. More generally, a modified analysis is required to account for the lack of information about the case-control status of the pseudo-siblings. The pseudo-siblings differ from the case only in their genotype at this locus, and hence it is not possible under this design to estimate the main effects of environmental exposures. However, interactions between genotype and environment can still be estimated. A drawback of this design is that it requires information from both parents of a case: this incurs greater cost and also assumes that both parents are living.

A special type of family matched case-control study is a twin study, in which twin pairs are identified for which one of the pair has the outcome under study, for example the occurrence of disease, and the other does not. Identical (monozygotic) twins are matched exactly on genetic factors. Twin studies are therefore very useful in studies of relationships between non-genetic exposures and outcomes that are believed to have a strong genetic component, enabling control for other environmental factors. Studies comparing non-identical (dizygotic) twins with identical twins can help to unravel the relative importance of genes and environmental factors on an outcome of interest.

6.7 Augmenting case-control data using family members

There is a different way of using family members in case-control studies. First, a standard case-control sample of say n_1 cases and n_0 controls is taken from the underlying population. Then one or more family members of each case and each control are also recruited to the study and the case or control status of each such family member is recorded. Covariate information is obtained for the original cases and controls and also for their recruited family members. The originally sampled cases and controls are sometimes referred to as *probands*. Family members of the cases and controls may have either outcome. The aim of studies of this type is to model the outcome status of family members, typically to gain information about the contribution of genetic factors to the outcome of interest, usually a disease.

One way to investigate exposure-outcome associations using case-control family data of this type is to view each family as a matched set. In each set there are individuals who are similar with respect to many genetic and environmental factors. Let Y_{ij} denote the outcome for the jth member of the ith family, and let X_{ij} denote their exposure or vector of exposures and covariates. As in earlier discussions, we suppose that the inverse model of interest is logistic:

$$\Pr{}^{\mathcal{I}}(Y_{ij} = y | X_{ij} = x_{ij}) = L_y(\alpha_i + \beta^T x_{ij}), \quad y = 0, 1.$$

The intercepts α_i allow for between-family differences in the baseline probabilities of an individual's having the outcome. It may be important to incorporate into the vector X some variables that indicate the individual's position within the family structure, in order to accommodate the possibility that an individual's exposures may depend on the disease status of his or her relatives. Extending slightly the results in Chapter 4 relating to matched case-control studies with one case per matched set, we have that the relevant conditional likelihood for the family data is

$$\prod_{i=1}^{n} \frac{\prod_{j \in F_{i1}} e^{\beta^T x_{ij}}}{\sum_{\mathcal{F}_{i1}} \prod_{k \in \mathcal{F}_{1i}} e^{\beta^T x_{ik}}}, \tag{6.38}$$

where F_{i1} denotes the set of cases in the ith family, which number n_{i1} say, and $\mathcal{F}_{i1}$ denotes the set of all possible sets of size n_{i1} of individuals in the ith family, that is, all possible configurations of n_{i1} cases within the family.

A drawback to the conditional analysis just described is that families in which the originally sampled individual is a control and all the recruited family members are also controls will be uninformative. In this circumstance an alternative unconditional analysis may be preferred, which we outline briefly here. For simplicity we now consider a family-based case-control study in which just one family member is recruited alongside each originally sampled case and control, the probands. Let Y_{i1} denote the case or control status of the original recruit and Y_{i2} that of the family member in the ith family, and let X_{i1}, X_{i2} denote their corresponding exposures, which are for simplicity assumed to be categorical. The full likelihood for the case-control family data is

$$\prod_i \Pr^{\mathcal{D}_1}(X_{i1} = x_{i1} | Y_{i1} = y_{i1})$$
$$\times \Pr^{\mathcal{D}_2}(Y_{i2} = y_{i2}, X_{i2} = x_{i2} | Y_{i1} = y_{i1}, X_{i1} = x_{i1}), \quad (6.39)$$

where $\mathcal{D}_1$ indicates the sampling model for the original case or control and $\mathcal{D}_2$ indicates the sampling model for family members of cases and controls. The likelihood closely resembles those discussed in connection with two-stage case-control sampling. Like those, it is subject to the constraints on the probability distribution of the original cases y_{i1} implied by their having being selected. It can be shown that this likelihood gives rise to the same estimates of logistic parameters of interest as would arise had the data been obtained prospectively, for example by the sampling of families at random from the underlying population. This uses the assumption that, conditioned on their own covariates, an individual's outcome status is independent of family members' covariates.

In the unconditional approach just described it is assumed that the family-specific effect on the outcome is totally accounted for in the covariates x_{ij}, which may in fact be implausible, especially if there are believed to be strong genetic factors at play. However, extensions are possible including a random effect for 'family'. This requires a parametric assumption about the distribution of the random effects.

Example 6.4: Use of family members as controls Multi-stage sampling and the use of family members as controls, both of which we have discussed in this chapter, can be combined, as we describe in this example. Whittemore and Halpern (1997) reported a three-stage study investigating risk factors for prostate cancer. At stage 1, prostate cancer cases and a set of non-diseased controls were sampled with frequency matching on ethnicity, age and region of residence. Lifestyle exposures were obtained within this case-control sample and the participants were also asked whether they had a father or brother with prostate cancer, thus dividing the cases and controls into those with and without a family history of the disease. At stage 2, non-random samples were taken within the four subgroups defined by case or control status and family history or no family history. Individuals sampled at this stage provided further information on their family history of prostate cancer, for example ages at diagnosis for fathers and brothers and family size. The stage-3 sample consisted of families identified at stage 2 in which three or more members have prostate cancer. Individuals at the third stage, including the family members, provided biological material for DNA analysis.

Notes

Section 6.2. Two-stage case-control designs with a cohort at the first stage, and more general possibilities, were considered in a series of papers by Scott and Wild, where many technical statistical details were given. See Scott and Wild (1986, 1991, 1997, 2001) and Wild (1991).

Section 6.2.5. Two-stage case-control sampling in which the second stage involves sampling within strata defined by both exposure and outcome was described by White (1982), who noted the possibility of a case-control sample at the first stage. White (1982) described the analysis for a binary exposure and a categorical covariate with case-control sampling at both stages. Methods for analysis were further developed by Breslow and Cain (1988). See Breslow and Chatterjee (1999) for an application of the two-stage case-control design and a summary of methods of analysis.

Sections 6.3.2. Likelihood-based methods for two-stage case-control studies are described in the papers of Scott and Wild. Extensions to continuous exposures are also discussed there. The methods are not restricted to a logistic model; see Wild (1991) for an application using the proportional odds model. Scott and Wild (1991) extended the results of Wild (1991) to the stratified sampling situation. The pseudo-likelihood approach was also described by Cosslett (1981) in the econometrics literature. See Schill and Drescher (1997) for an overview of methods of analysis for two-stage studies. See also Hsieh *et al.* (1985).

Section 6.3.3. A pseudo-likelihood analysis of two-stage case-control studies with a case-control sample at both stages was outlined by Breslow and Cain (1988). See also Fears and Brown (1986) and Breslow and Zhao (1988). A commentary was given by Cain and Breslow (1988). Breslow and Holubkov (1997) described the full maximum-likelihood procedure for two-stage case-control studies with a case-control sample at both stages, including extensions to continuous covariates.

An alternative method for the analysis of two-stage case-control data is a weighted-likelihood approach. See appendix section A.7 for a brief general outline of this kind of approach. For a discussion of weighted likelihood for two-stage case-control studies see Flanders and Greenland (1991), Wild (1991) and Schill and Drescher (1997). For comparisons between full-likelihood, pseudo-likelihood and weighted-likelihood analyses for two-stage case-control studies see Wild (1991), Breslow and Holubkov (1997), Scott and Wild (1986, 2002) and Breslow and Chatterjee (1999). For the use of a weighted approach in one-stage case-control studies see Scott and Wild (2002).

Section 6.4.1. Optimal sampling strategies for improvement of the estimation of interaction terms in the context of gene-environment interactions were discussed by Saunders and Barrett (2004) and Sturmer and Brenner (2002). Chatterjee and Carroll (2005) described an efficient method of analysis that exploits an assumption of gene-environment independence.

Section 6.4.2. Breslow and Cain (1988) presented some efficiency results, comparing a two-stage case-control study with a one-stage study. They also compared balanced with optimal two-stage sampling. See also Hanley *et al.* (2005). A method for optimal sampling at the second stage using Neyman-optimized allocation is outlined in Whittemore and Halpern (1997).

Section 6.4.3. Extensions of the full-likelihood analyses for two-stage studies, with either prospective or case-control sampling at the first stage, to general multi-stage designs were presented by Lee *et al.* (2010), who also gave a useful account of previous results. Whittemore and Halpern (1997) discussed the analysis of multi-stage sampling designs in a more general context, including the use of weighted analyses.

Section 6.5. Langholz and Goldstein (2001) described a number of different designs for special sampling in matched case-control studies. See also Langholz and Goldstein (1996), which described a two-stage sampling design for nested case-control studies. These may be considered as a special type of matched study.

Section 6.6.2. Gauderman *et al.* (1999) gave an overview of family-based case-control designs using relatives as controls. See also Maraganore (2005) for a non-statistical overview with a focus on studies in neurology. Witte *et al.* (1999), Weinberg and Umbach (2000) and Schaid (1999), among many others, discussed the use of family members as controls in studies of gene-environment interactions in epidemiology. Austin and Flanders (2003) described special methods of analysis for studies that use the children of cases as controls. See Witte *et al.* (1999) for the modified analysis for the pseudo-sibling design.

Section 6.7. Whittemore and Halpern (2003) outlined the conditional analysis for this type of family-based study, as well as other possibilities. Whittemore

(1995) established the relationship between the retrospective and prospective likelihoods for an unconditional analysis of family-based case-control data. See also Zhao *et al.* (1998) and Whittemore and Halpern (2003) for a comparison with the conditional approach. Neuhaus *et al.* (2002) presented some extensions to incorporate a wider class of family-based designs, specifically designs in which families are sampled on the basis of having a particular pattern of outcomes within the family. Further extensions to family-specific models are described in Neuhaus *et al.* (2006). See also Wang and Hanfelt (2009). Chatterjee and Carroll (2005) described an alternative conditional analysis for family-matched case-control studies that exploits an assumption that genotype and environmental exposure are independent within families. Siegmund and Langholz (2001) discussed multi-stage family-based case-control-type designs for assessing gene-environment interactions.

7

Nested case-control studies

- The nested case-control design accommodates case event times into the sampling of controls.
- In this design one or more controls is or are selected for each case from the risk set at the time at which the case event occurs. Controls may also be matched to cases on selected variables.
- Nested case-control studies are particularly suited for use within large prospective cohorts, when it is desirable to process exposure information only for cases and a subset of non-cases.
- The analysis of nested case-control studies uses a proportional hazards model and a modification to the partial likelihood used in full-cohort studies, giving estimates of hazard ratios. Extensions to other survival models are possible.
- In the standard design, controls are selected randomly from the risk set for each case; however, more elaborate sampling procedures for controls, such as counter-matching, may gain efficiency. A weighted partial-likelihood analysis is needed to accommodate non-random sampling.

7.1 Preliminaries

We have focused primarily so far on case-control studies in which the cases and controls are sampled from groups of individuals who respectively do and do not have the outcome of interest occurring within a relevant time window. This time window is typically relatively short. If cases are to be defined as those experiencing an event or outcome of interest occurring over a longer time period, or if the rate of occurrence of the event is high, then the choice of a suitable control group requires special care. The idea of introducing *time* into case-control sampling was discussed in Section 1.4. In Section 2.10 we outlined the sampling of controls within short time intervals, and in Section 3.10 the possibility was noted of extending this to continuous time with controls selected for each case at its event time. In the present chapter we focus on elaborating the latter idea, of case-control

160

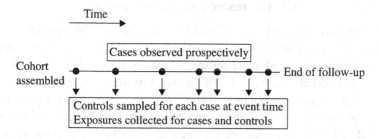

Figure 7.1 Outline of a nested case-control study. The arrows indicate that controls are selected for each case at its event time.

sampling in continuous time, which we refer to as *nested case-control sampling*.

The nested case-control design is an extension of a case-control study to a survival analysis setting in which the outcome of interest is an event with an associated event time and the focus is on making inferences about whether hazard rates are associated with exposures of interest. Case-control studies in which the event times are a key part of the sampling process are sometimes referred to as *synthetic retrospective studies*. They may be more appropriately viewed as a special form of prospective study. Another design of this type is the *case-subcohort design*, which is considered in Chapter 8.

The nested case-control design is appropriate in two main situations.

(1) The first situation arises when we wish to take a case-control-type sample within a large assembled cohort followed over time. Prospective cohorts arise commonly in epidemiological studies, for example, and it is often desirable to reduce costs by obtaining particular exposure measurements only in a subset of the cohort.

(2) The second situation arises when we wish to select a population-based case-control sample over a fairly long period of time. In this situation the underlying population is not an assembled cohort of individuals participating in a study but a well-defined but possibly varying population within which we can observe when individuals have the event of interest. An example is when a case-control sample is to be taken from a country-wide health register over a period of several years or more.

We focus mainly on the first situation in our descriptions, as illustrated in Figure 7.1, though the essence is the same in both.

7.2 The nested case-control design

Consider a prospective cohort of individuals followed up for an outcome of interest. The cases are those individuals who have the event of interest during the follow-up period. In a later section we discuss briefly the possibility of recurrent events. Individuals who do not have the event of interest will have a right censored event time; that is, we observe the time up to which it is known they have not had the event. Individuals may be censored before the end of follow-up by leaving the population, and it is also possible that there is an influx into the population.

Figure 7.2 illustrates the underlying cohort within which cases occur and the procedure for selecting a nested case-control sample within that cohort. The main steps are as follows.

(1) Cases are identified within the cohort at the time at which they are observed to experience the event of interest. Often all cases observed during a particular period of follow-up will be studied.

(2) At a given event time the *risk set* is the set of individuals who were eligible to experience the event at that time, that is, who still remain in the cohort, have not yet experienced the event just prior to the observed event time and have not been censored.

(3) We identify the risk set at each case's event time and take a sample of one or more individuals from that risk set. We refer to these individuals as the controls for that case; a key feature of the nested case-control design is that individuals can be sampled as controls for more than one case and moreover that individuals sampled as controls may subsequently become cases.

There are a number of ways in which individuals can be sampled to form the control set at each event time. Under the standard nested case-control design, at each event time the controls are selected by simple random sampling from the risk set, *excluding* the case itself. The number of controls sampled at each event time is usually small and commonly there is just one control per case. As an extension to the above, at each event time the controls may be sampled from a subset of the risk set so as to match the case also on selected variables.

Full information on exposure variables and other variables of interest is obtained within the nested case-control sample. In nested case-control studies within prospective cohorts, often exposure information will have been obtained prior to the outcome, for example at recruitment to the

(a) Using time since study recruitment as the time scale.

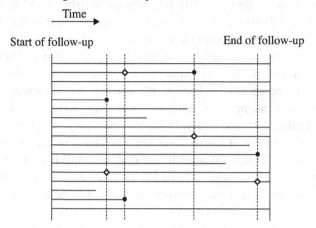

(b) Using age as the time scale.

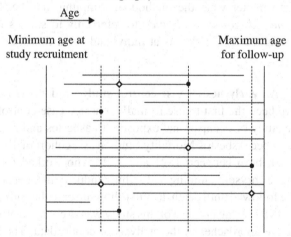

Figure 7.2 Sampling a nested case-control study with one control per case, using time since study recruitment or age as the time scale. The solid lines represent the time period over which individuals are observed. Cases occur over the course of follow-up (•), individuals may leave the population or may survive to the end of the follow-up period or maximum age of observation. The dotted lines pass through members of the risk set at each event time. One control (◇) is selected for each case from its risk set.

underlying cohort study, and perhaps periodically thereafter for time-varying exposures. Some of this information will be cheap to observe for the whole cohort. However, for some exposures the material required to obtain an exposure measurement may be collected and stored, but the actual measurement required for a statistical analysis is not obtained because of cost. For example, blood samples may be obtained and stored for later use. The prospective collection of exposure and covariate information eliminates some potential sources of bias that could arise if exposures were ascertained retrospectively. However, there may be other situations in which an exposure of interest was not considered in earlier rounds of data collection, perhaps because it has only recently been found to be of potential interest in other related studies. In this case some prospective studies may obtain data retrospectively within nested case-control samples.

The term *nested case-control study* is sometimes used also to refer to any case-control study in which the underlying population is an assembled cohort, no matter what the method of sampling may be. Here we use the term *nested case-control study* to refer only to studies in which controls are sampled from risk sets at individual event times, as outlined above.

Example 7.1: An early nested case-control study   Liddell *et al.* (1977) appear to have been the first to use formally a nested case-control design. This was in a study of occupational exposure to asbestos and risk of lung cancer. The Quebec Asbestos Mortality Study was a cohort of 10 951 men and 440 women born between 1891 and 1920 who worked for at least one month in the chrysotile mining and milling industry in Quebec. Cohort members were followed until death, loss to follow-up or to the end of follow-up at the end of 1973. One purpose of this study was to compare prospective and retrospective approaches to the analysis of cohort data. The analyses focused just on the males in the cohort. Up to the end of follow-up there were 215 lung cancer deaths. The authors noted the matched case-control design and stated that 'As there is no fundamental reason why subsequent death should disqualify a control, the approach could be extended further to use all living [members of the cohort] as controls for each man dying, or more realistically for each age at death'. Taking this approach, five controls were selected for each case by random sampling from those born in the same year as the case and who were known to have survived at least until the year after that in which the case died.

7.3 Analysis of nested case-control studies

The analysis of data arising from a nested case-control study derives from the analysis of a prospective cohort study in which we have collected exposure and other covariate data and observed either a survival time, or 'event time', or a censoring time for all those in the cohort. We will focus on using the semi-parametric proportional hazards model, which is a commonly used method of analysis for studies investigating associations between exposure variables and events of interest and is used particularly extensively in epidemiology.

Let t denote the time which has elapsed since an individual became eligible for study. This is usually defined as the time since recruitment to the study, or sometimes as age or as the time since a relevant critical event. In general the time origin for each individual should be chosen to put different individuals with the same background variables on an equal footing, that is, having as nearly as possible equal probabilities of becoming a case. Under the proportional hazards model the hazard function at time t for an individual with vector of possibly time-dependent exposures and covariates $x(t) = \{x_1(t), \ldots, x_p(t)\}^T$ is

$$h\{t; x(t)\} = h_0(t) \exp\{\beta^T x(t)\}, \tag{7.1}$$

where $h_0(t)$ is the hazard for an individual with baseline covariates and $\beta = (\beta_1, \ldots, \beta_p)^T$ is a vector of regression coefficients to be estimated. For a binary exposure, β_k is the log hazard ratio, or log rate ratio, for an exposed individual relative to an unexposed individual, taken conditionally on the other covariates. For a continuous exposure, β_k is the log hazard ratio associated with a unit increase of exposure, the other covariates being held fixed.

In a nested case-control study, usually all observed cases from the underlying cohort will be included. We denote the ordered event times for the cases as $t_1 < t_2 < \cdots$. In practice, some event times may be tied, that is, simultaneous, but we omit this complication. The case at time t_j is indexed by i_j. The set of individuals in the full risk set at time t_j is denoted R_j. Note that R_j includes the case itself.

To complete the specification, the method for choosing controls also needs to be formulated. At the failure time t_j we let $\tilde{R}_j$ be the set of individuals selected at time t_j, including the case. We refer to this as the *sampled risk set*. In the scheme illustrated in Figure 7.2, with one control sampled for each case, $\tilde{R}_j$ denotes the case and the single control chosen at time t_j.

At the failure time t_j we let $\mathcal{H}_j$ denote all the information on individuals at risk in the cohort up to just before time t_j, including information on failures, censorings and exposures. Some or all exposures will be observed only in individuals sampled to the nested case-control study. Now define $\tilde{\mathcal{H}}_j$ to be a version of $\mathcal{H}_j$ extended to include information on which individuals have been sampled to the nested case-control study just before time t_j.

We now define the *design probability*, $p(\tilde{R}_j; i_j, \tilde{\mathcal{H}}_j)$, as the probability of $\tilde{R}_j$. In other words $p(\tilde{R}_j; i_j, \tilde{\mathcal{H}}_j)$ is the probability of observing that particular sampled risk set given the identity i_j of the case at t_j and given all the history summarized in $\tilde{\mathcal{H}}_j$. In particular the probability $p(\tilde{R}_j; i_j, \tilde{\mathcal{H}}_j)$ will assign probability unity to the choice of i_j: the case was guaranteed to be in the sampled risk set. We assume that the hazard for any individual, given their history, and in particular their values of x, has the proportional hazards form (7.1).

We define a partial likelihood by taking a contribution from the sampled risk set $\tilde{R}_j$ at each time t_j. This contribution consists of the probability that, given that exactly one event of interest (the case) occurs at the time t_j and given that the sampled risk set is the one observed, the case is in fact individual i_j and not one of the other individuals in $\tilde{R}_j$. This gives the contribution

$$\frac{p(\tilde{R}_j; i_j, \tilde{\mathcal{H}}_j) \exp\{\beta^T x_{i_j}(t_j)\}}{\sum_{k \in \tilde{R}_j} p(\tilde{R}_j; k, \tilde{\mathcal{H}}_j) \exp\{\beta^T x_k(t_j)\}}. \tag{7.2}$$

The terms contributing to the denominator are proportional to the probabilities that the case at time t_j is individual k given the history $\tilde{\mathcal{H}}_j$ and given that the sampled risk set is that observed. That individual k is the case is contrary to fact unless $k = i_j$. The full partial likelihood is now obtained by taking the product of (7.2) over all times t_j at which cases arise.

A special situation occurs when complete exposure and covariate data are available on the whole cohort. Then $p(\cdot)$ assigns probability unity to all members of the cohort at risk at the time in question, and the contribution of the factors $p(\cdot)$ can be ignored. That is, the partial likelihood takes the simple form

$$\prod_j \frac{\exp\{\beta^T x_{i_j}(t_j)\}}{\sum_{k \in \tilde{R}_j} \exp\{\beta^T x_k(t_j)\}}, \tag{7.3}$$

where the product is over all times t_j and here $\tilde{R}_j = R_j$. This is the familiar partial likelihood for the analysis of a full cohort, using the semi-parametric proportional hazards model.

The simplest possibility in a nested case-control study is that, as each case arises, a sample of $m - 1$ controls is selected at random without replacement from $R_j \backslash i_j$, that is, from the risk set excluding the case itself. Then the $p(\cdot)$ are all equal to

$$\binom{n_j - 1}{m - 1}^{-1}, \tag{7.4}$$

where n_j is the number of individuals in the full risk set at time t_j. Hence the probabilities $p(\cdot)$ cancel in (7.2) and the simpler form (7.3) again emerges. The reason for this is that, under this scheme, the sampling of control individuals from the risk set at a given time does not depend on the actual identity or other characteristics of the case.

The simple analysis using (7.3) that arises under the above random sampling schemes can also be seen to hold when the number of controls per case varies, that is, if we replace m by m_j, say. There are other sampling procedures, considered below, under which the probabilities $p(\cdot)$ *do* depend on the identity of the case in the set, and the simple form of analysis is not applicable.

7.4 Further comments on analysis

7.4.1 Connection with conditional logistic regression

The nested case-control study is an extended form of matched case-control study, because the data comprise a series of case-control sets matched on time and possibly on other features too. In the analysis each case is compared with its own control set in a conditional analysis. The analysis based on the partial likelihood in (7.3) is identical to the conditional logistic regression analysis for standard matched case-control studies.

7.4.2 Large-sample properties

The large-sample properties of the partial likelihood for a nested case-control study are dependent on the successive contributions to the score functions being uncorrelated after conditioning on the past. This is the case only if $\tilde{\mathcal{H}}_j$ is nested within $\tilde{\mathcal{H}}_k$ for $j < k$, which is true provided that the risk set sampling at each failure time is independent of the risk set sampling at all earlier failure times, which is crucial for the validity of a nested case-control analysis using a weighted partial likelihood and is discussed briefly in the appendix.

7.4.3 Efficiency relative to a full cohort analysis

In this section we consider the asymptotic efficiency of a nested case-control study, as described above, relative to the corresponding full cohort analysis that would be performed if exposure and covariate information was available for all individuals. For a single exposure of interest the efficiency is

$$\left\{ E\left(-\frac{\partial^2 \tilde{l}}{\partial \beta^2}\right)\right\}^{-1} E\left(-\frac{\partial^2 l}{\partial \beta^2}\right), \tag{7.5}$$

where $\tilde{l}(\beta)$ and $l(\beta)$ are respectively the log partial likelihoods from the nested case-control study and the corresponding full cohort study. The expected information from the nested case-control study with random sampling of controls is

$$E\left(-\frac{\partial^2 \tilde{l}}{\partial \beta^2}\right) = E\left(\sum_{t_j} \frac{s_2 s_0 - s_1^2}{s_0^2}\right), \tag{7.6}$$

where the sum is over all failure times t_j, as was the product in (7.3). Here

$$s_r = \sum_{k \in \tilde{R}_j} \{x_k(t_j)\}^r e^{\beta x_k(t_j)}. \tag{7.7}$$

We focus on a binary exposure and let

$$p_1(t_j) = \Pr(X_k(t_j) = 1 | k \in R_j)$$

denote the probability of being exposed at time t_j for an individual in the risk set at t_j; we define $p_0(t_j) = 1 - p_1(t_j)$.

The probability that the case i_j at time t_j is exposed, conditioned on its being in the risk set at t_j, is thus

$$\Pr\left(X_{i_j}(t_j) = 1 | i_j \in R_j, Y_{i_j}(t_j) = 1\right) = \frac{h_0(t_j)e^\beta p_1(t_j)}{h(t_j)},$$

where $Y_k(t_j)$ takes the value 1 if individual k is a case at time t_j and the value 0 otherwise, and $h(t) = \Sigma_x h_0(t)e^{\beta x} p_x(t)$.

By conditioning on the exposure status of the case at each failure time it can be shown, using the above results, that the expected information in

(7.6) can be written as

$$E \sum_{t_j} \sum_{s=1}^{m-1} \binom{m-1}{s} \frac{h_0(t_j)}{h(t_j)} \frac{s(m-s)e^\beta}{(se^\beta + m - s)^2} p_1(t_j)^s p_0(t_j)^{m-s}$$

$$= \int \sum_{s=1}^{m-1} \binom{m-1}{s} p_0(t) \frac{se^\beta}{se^\beta + m - s}$$

$$\times p_1(t)^s p_0(t)^{m-1-s} h_0(t) \tilde{S}(t) dt, \tag{7.8}$$

where $\tilde{S}(t)$ is the probability of an individual's remaining on-study (that is, under observation) and event-free beyond time t, and the integral is over the follow-up time period. If there is no censoring except by the outcome of interest then $\tilde{S}(t) = S(t)$, the survivor function at time t, that is, the probability of survival beyond time t.

The full cohort analysis can be thought of as a nested case-control analysis with a large number of controls at each failure time. The expected information from the full cohort analysis can therefore be derived by finding the limit of $E(-\partial^2 \tilde{l}/\partial\beta^2)$ as m tends to infinity. By noting that $m^{-1} \sum_{k \in \tilde{R}_j} x_k(t_j) \to p_1(t_j)$ as $m \to \infty$, it follows that

$$E\left(-\frac{\partial^2 l}{\partial\beta^2}\right) = \int \frac{p_1(t)p_0(t)e^\beta}{p_1(t)e^\beta + p_0(t)} h_0(t)\tilde{S}(t)\, dt. \tag{7.9}$$

It can be shown directly from these formulae that if $\beta = 0$ then the information achieved with $m - 1$ controls is $(m - 1)/m$ times that using the full cohort. This corresponds to comparisons of the variance of the difference between two sample means, one $m - 1$ times the size of the other. If, however, $\beta \neq 0$ then the evidence supplied by a case is relatively increased or decreased depending on the sign of β.

7.5 Some extensions

It is very common in a nested case-control study to match cases to controls to ensure that all controls have the same value of specific features W as the corresponding case; age and gender are typical examples. This mimics the use of matching in a standard case-control study and is used to control for those variables whose main effect on the outcome we do not wish to estimate and which are in some sense prior to the explanatory variables of concern. The sampling of a nested case-control study from an assembled cohort means that many covariates will be known for the full cohort, making additional matching straightforward. At each failure time, controls are sampled randomly within strata defined by the matching variables. Then if the design probability $p(\cdot)$ assigns equal probability to all relevant individuals having the required values of W and zero probability otherwise, again cancellation occurs in (7.2). The baseline hazard in model (7.1) is now defined within strata. Note also that if the hazard contains an arbitrary additional factor depending only on W, that too will cancel from the resulting partial likelihood.

The sampling of controls from risk sets under the nested case-control design makes this design suitable for use also in studies of recurrent events such as an infection from which people recover. In this situation an individual who has had an occurrence of the event of interest may be returned to the risk set at some appropriate time after the occurrence, for example after an infection may reasonably be assumed to have cleared. Exposure information measured fairly frequently over time would usually be required in this setting.

Example 7.2: A nested case-control study with matching Nested case-control samples within the Nurses' Health Study and the Health Professionals Follow-up Study, both large US cohort studies, were used to investigate associations between inflammatory markers and the risk of coronary heart disease by Pai *et al.* (2004). Of the 121 700 Nurses' Health Study participants, 32 826 provided a blood sample upon request and of 51 529 Health Professionals Study participants 18 225 did so. Female cases were those who had a non-fatal myocardial infarction or fatal coronary heart disease between the date of the blood sample and June 1998, and male cases were those who had such an event between the date of the blood sample and 2000. The use of a nested case-control design meant that biological measures were needed only for a subset of individuals.

There were 249 female cases and 266 male cases of myocardial infarction. At each case's time of occurrence two controls were randomly sampled from the risk set, excluding the case, to match the case on age, smoking status, date of blood sample and, in the male study only, fasting status at the time of the blood sample.

7.6 More elaborate sampling procedures

7.6.1 Preliminaries

The previous discussion centred on simple random sampling from the risk set for the choice of controls for each case. It is helpful to consider more elaborate sampling schemes intended to improve precision. For each it is important to consider whether modification of the simple partial likelihood in (7.3) is required.

7.6.2 Counter-matching

The first such special control sampling scheme that we consider is *counter-matching*, which uses covariate information from the full cohort to influence the sampling of controls at each event time in such a way that the variability of the exposure is maximized within $\tilde{R}_j$, as far as is possible using the information from the full cohort. There are two main situations in which counter matching may be used.

(1) *Counter-matching on a surrogate exposure.* First suppose that the main exposure of interest is expensive or difficult to obtain and is going to be available only in the nested case-control sample. However, a cheaper but error-prone measurement of exposure may be available in the full cohort. We refer to this as *surrogate exposure*.

Consider for simplicity a binary exposure and suppose that at event time t_j we wish to sample $2\tilde{m} - 1$ controls from the risk set excluding the case, $R_j \backslash i_j$. Under the counter-matching design $\tilde{m} - 1$ controls are sampled from the part of $R_j \backslash i_j$ with the *same* surrogate exposure as the case, and $\tilde{m}$ controls are sampled with the *opposite* surrogate exposure to the case. This is illustrated in Figure 7.3. The 'expensive' exposure measurement, along with any other variables of interest, is then obtained for the nested case-control sample. Because the sampling probabilities are different depending on the surrogate exposure of the case, the design probabilities in (7.2) do not cancel; the modification required is developed below.

(2) *Counter-matching on exposure.* The second situation occurs when the main exposure is available for the full cohort but there are one or more important confounders that can be measured only in a nested case-control sample. The sampling procedure is as in (1) except that it is based on the real rather than the surrogate exposure. All confounder information is then obtained for individuals in the nested case-control sample.

Counter-matching can be extended to accommodate exposures with more than two categories by selecting controls for each case in such a way that each sampled risk set contains a fixed number of individuals from each exposure category. For continuous exposures the range of the exposure

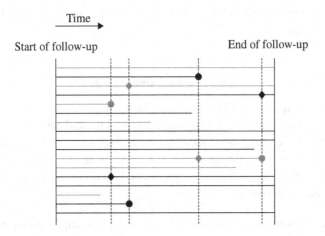

Figure 7.3 Sampling a counter-matched (on surrogate exposure) nested case-control study with one control per case. Individuals are classified according to their surrogate exposure (black or grey). The dotted lines pass through individuals in the risk set for each case. For cases with black exposure (●) a control with grey exposure (◆) is sampled from the risk set, and vice versa.

must be divided into groups and counter-matching performed within these groups.

In a counter-matched design the sampling of controls at each failure time is dependent on the case's exposure and hence on the identity of the case. Consider an exposure with L categories. At event time t_j let $n_l(t_j)$ denote the number of individuals in the risk set R_j who are in stratum l ($l = 1, \ldots, L$) of the exposure or surrogate exposure, depending on whether we are in situation (1) or (2). Let $\tilde{m}_l$ denote the number of individuals, including the case itself, to be sampled from each stratum l ($l = 1, \ldots, L$) of the risk set R_j at each failure time. The stratum of the case at time t_j is l_{i_j}. The design probability is

$$p(\tilde{R}_j; i_j, \tilde{\mathcal{H}}_j) = \left\{ \prod_{l \setminus l_{i_j}} \binom{n_l(t_j)}{\tilde{m}_l} \binom{n_{l_{i_j}}(t_j) - 1}{\tilde{m}_{l_{i_j}} - 1} \right\}^{-1}$$

$$= \left\{ \frac{\tilde{m}_{l_{i_j}}}{n_{l_{i_j}}(t_j)} \prod_l \binom{n_l(t_j)}{\tilde{m}_l} \right\}^{-1}, \qquad (7.10)$$

where $m_{l_{i_j}}$ is the number of individuals in $\tilde{R}_j$ who are sampled from the same stratum as the case i_j at time t_j, including the case itself, and $n_{l_{i_j}}(t_j)$ is the number of individuals at risk at t_j who are in the same exposure stratum as the case. The modified partial likelihood (7.2) that is used to analyse nested case-control data with controls sampled using counter-matching is therefore, after some simplification, the product over event times t_j of factors

$$\frac{\exp\{\beta^T x_{i_j}(t_j)\}\, n_{l_{i_j}}(t_j)/m_{l_{i_j}}}{\sum_{k\in\tilde{R}_j} \exp\{\beta^T x_k(t_j)\}\, n_{l_k}(t_j)/m_{l_k}}. \tag{7.11}$$

The weights are inversely proportional to the probability of selection of that configuration and, as such, induce an unbiased estimate of the required formal likelihood function.

To assess the asymptotic efficiency of the counter-matching designs relative to a nested case-control study with random sampling of controls involves calculations similar to but more detailed than those for the simpler form of nested case-control design. The efficiency depends on a number of factors, including the distribution of the exposure in the underlying population and the predictive probability of any surrogate exposure. We refer to the notes for details.

Example 7.3: Counter-matching in a study of gene-environment interactions Bernstein *et al.* (2004) described the use of counter-matching to study potential interactions between gene variants that predispose women to breast cancer and exposure to ionizing radiation. The outcome of interest was a second breast cancer among women previously treated for breast cancer. The population underlying this case-control study was five population-based cancer registries. The cases were 700 women identified from the registries as having had two separate primary breast cancer diagnoses, the first occurring between 1985 and 2000 and the second occurring at least one year after the first. All were women aged under 55 at the time of the first diagnosis who had remained alive, so they could be contacted for this study.

For each case two controls were sampled from the registries; the controls were the same age as the case (within five years), had received their first breast cancer diagnosis within four years of the case and had survived without any new cancer diagnosis up until the time when the case had her case-defining second diagnosis. Some other criteria for case and control selection are described in the paper.

The selection of controls used also counter-matching based on radiation exposure. A binary indicator of whether women had received ionizing radiation treatment for their first breast cancer was available from the cancer registry records. Each case who had received radiation treatment was matched to one control who had also received radiation treatment and one who had not. Each case who had not received radiation treatment was matched to two controls who had received radiation treatment. The binary indicator of radiation treatment from the records was used as a surrogate for the actual exposure of interest which was radiation dose; this was estimated among all cases and controls using a range of information about treatment and patient characteristics.

Women selected for the case-control study provided blood samples, which were used to determine genotypes for the genes of interest. They also provided information on potential confounders.

7.6.3 *Quota sampling*

A different scheme, quota sampling, may be useful when there is no surrogate, the exposure can be measured only in the nested case-control sample and, moreover, one or more of the exposure categories is rare. Under this design each exposure category is assigned a quota; that is, each matched set contains at least a specified number of individuals in each exposure group. For each case, controls are sampled sequentially until all quotas are met. For example, for a binary exposure the sampling of controls continues until the sampled risk set contains at least m_1 exposed individuals and at least m_0 unexposed individuals, resulting in totals of say $\tilde{m}_1$ and $\tilde{m}_0$ exposed and unexposed individuals in the sampled risk set. Here, either $\tilde{m}_1 = m_1$ or $\tilde{m}_0 = m_0$, depending on which total is achieved first in the sampling process.

Because the sampling scheme depends on the status of the case, weighting of the partial likelihood is needed, as in (7.2). The design probabilities in this situation come from a negative multi-hypergeometric distribution. We shall not give details of the calculations involved. For quota sampling based on a binary exposure $X(t)$, the analysis uses the weighted partial likelihood

$$\prod_j \frac{\exp\{\beta x_{i_j}(t_j)\} \sum_l \{(m_l - x_{i_j}(t_j))I(\tilde{m}_l = m_l)\}}{\sum_{k \in \tilde{R}_j} \exp\{\beta x_k(t_j)\} \sum_l \{(m_l - x_k(t_j))I(\tilde{m}_l = m_l)\}}, \qquad (7.12)$$

where $I(\tilde{m}_l = m_l)$ is an indicator taking the value 1 if $\tilde{m}_l = m_l$ and the value 0 otherwise and the product is over event times t_j.

The number of individuals needed to produce the desired quotas of individuals in each set can be predicted only in some average sense. However, an obvious more serious practical drawback of this design is the possibility that some cases might demand study of a large number of potential controls before the required discordance is achieved. This suggests setting an upper limit on the number of controls per case. Nevertheless, when quota sampling is practicable the gains in efficiency would appear to be appreciable; we have been unable find examples of the use of the method.

7.7 Further comments on the sampling of controls

One possibility sometimes considered, but which in fact is usually a bad idea, is to postpone the choice of controls until the end of the period of study and to take only 'real' controls, that is, individuals who have survived the whole period of study without becoming a case. This excludes, in particular, individuals selected as contemporaneous controls who become a case shortly thereafter. Some arguments against this were discussed in Section 1.4.

The bias induced by the above approach can be investigated mathematically. In Section 3.10 we introduced the 'instantaneous' odds ratio. If controls are sampled at random from the risk set at each failure time then this instantaneous odds ratio is equal to the hazard ratio. Consider a particular failure time t. For a single binary exposure we can write the hazard ratio at t as

$$\Lambda(t) = \frac{f(t;1)/f(t;0)}{S(t;1)/S(t;0)}, \tag{7.13}$$

where $f(t;x)$ denotes the probability density function of failure at time t for an individual with exposure x, and $S(t;x)$ is the corresponding survivor function. Suppose now that instead of sampling controls from the risk set at time t, we restrict control selection to those who survive up to some fixed time τ, which may for example be the time of the end of follow-up. For simplicity we assume no censoring except by the end of follow-up. Under this scheme the hazard ratio at t would be estimated as

$$\Lambda^*(t) = \frac{f(t;1)/f(t;0)}{S(\tau-t;1)/S(\tau-t;0)}. \tag{7.14}$$

To illustrate the potential difference between $\Lambda(t)$ and $\Lambda^*(t)$, suppose that the events under study occur according to a Poisson process, so that

$$f(t;x) = \lambda e^{\beta x} \exp(-\lambda t e^{\beta x}),$$
$$S(t;x) = \exp(-\lambda t e^{\beta x}).$$

This gives $\Lambda(t) = e^\beta$ and

$$\Lambda^*(t) = e^\beta \exp\{\lambda\tau(e^\beta - 1)\}.$$

Hence the sampling of controls only from the set of individuals who survive to time τ incurs a bias factor $\exp\{\lambda\tau(e^\beta - 1)\}$ in the hazard ratio estimate. In this special case, the bias factor will be close to unity if the baseline hazard rate λ is small or if β is close to zero, that is, if the exposure is only weakly associated with the outcome.

Other related possibilities to be avoided are that controls who subsequently develop some other outcome that may be related to the exposure are excluded or that those eligible for selection as controls at a given event time are required to remain free of the outcome for a fixed period post selection.

For some types of event there may be uncertainty about event times. For instance, if the event of interest is a cancer diagnosis then potential difficulties arise because of the long period over which an individual may have the cancer before they receive their diagnosis. Then one might question whether it would be appropriate to select as a control an individual who receives the same diagnosis at, say, one month after the case. It may be justified to restrict the potential control group to those not receiving the same diagnosis within, say, one year of the case; the time period to be allowed would depend on knowledge of the underlying condition. Of course, for outcomes that are rare within the underlying cohort it would be unlikely for an individual who received the same diagnosis shortly afterwards to be selected as a control. This issue applies equally in full cohort analyses.

7.8 Some alternative models

7.8.1 Preliminaries

The discussion in this chapter has so far hinged on the semi-parametric proportional hazards model (7.1). A reason for using a hazards-based formulation is that its definition as essentially a transition rate given a current state lends itself directly to the calculation of a partial likelihood based on such transitions. Use of the proportional hazards model is, however, by no means the only possibility. In this section we outline briefly three alternatives:

- a semi-parametric additive hazard model,
- a fully parametric hazard model,
- Aalen's linear hazard model.

7.8.2 An additive hazard model

If we are to exploit the cancellation of the baseline hazard from the partial likelihood the main modification arising is to replace the factor $\exp\{\beta^T x(t)\}$ in the proportional hazards model by some other function. The additive form is probably the most appealing, leading to an analysis based on a hazard

$$h_0(t)\{1 + \gamma^T x(t)\}. \tag{7.15}$$

The analysis is still performed using the partial likelihood. An empirical choice between this and the proportional hazards model (7.1), or possibly other forms, may be possible. Sometimes the choice between alternative forms is not critical to interpretation, in the light of the following property of alternative models both involving dependence on a linear combination of explanatory variables. In the above situation consider two components of x, say x_1 and x_2. Then the estimates from models (7.1) and (7.15) will be such that, approximately,

$$\hat{\beta}_2/\hat{\beta}_1 = \hat{\gamma}_2/\hat{\gamma}_1. \tag{7.16}$$

The qualitative explanation of this is that the ratio in (7.16) is the change in x_1 needed to induce the same effect as a unit change in x_2. This suggests that the ratio of regression coefficients is an indicator of the relative importance of the two explanatory variables that is not strongly dependent on the particular form of response variable. This has the important implication that assessment of the relative importance of two explanatory variables is often relatively insensitive to this aspect of model choice.

Similar remarks apply to the testing of a null hypothesis of zero effect. In any such specification appropriate definition of the explanatory vector x is important, especially if it is time dependent. The hazards-based specification is essentially that becoming a case is a Markov process with transition probability defined by the current state, that is, by the current value of x, which must therefore incorporate any relevant past information as part of its definition.

Example 7.4: Additive hazard model Darby *et al.* (2013) used a population-based nested case-control study to investigate whether women who had received radiation treatment for breast cancer were at increased risk of ischaemic heart disease. The underlying population was women in two cancer registries who had been diagnosed with breast cancer and had received radiotherapy: the Swedish National Cancer Registry (diagnoses between 1958–2001) and the Danish Breast Cancer Cooperative Group (diagnoses between 1977–2000). The cases were women who subsequently had a major coronary event, and the time scale was time since breast cancer diagnosis. For each case, one (Sweden) or two (Denmark) controls were selected from the underlying population from women who had survived as long as the case since their breast cancer diagnosis without a major coronary event, cancer recurrence, or death. The cases and controls were matched on country, age at breast cancer diagnosis and year of diagnosis within five years. Hospital records were used to estimate the radiotherapy dose for each woman in the nested case-control sample. An additive hazard model was used; there was evidence that this is the most appropriate for the particular exposure.

7.8.3 Fully parametric models

The semi-parametric formulation considered so far may be replaced by a fully parametric form in which the baseline hazard $h_0(t)$ is regarded as an unknown constant or some simple parametric function of time. Consider first a full-cohort situation and a parametric proportional hazards model in which the hazard function at time t for an individual with covariates $x(t)$ is $h(t; x(t))$. For example, for a parametric proportional hazards model with constant baseline hazard ρ, $h(t; x(t)) = \rho e^{\beta^T x(t)}$. In general, the contribution to the log likelihood from individual i is

$$l_i = Y_i \log h(\tilde{t}_i; x(\tilde{t}_i)) - \int_0^{\tilde{t}_i} h(s; x(s)) \, ds \qquad (7.17)$$

where $\tilde{t}_i$ denotes the observed failure or censoring time for individual i and Y_i is an indicator taking the value 1 if the ith individual has the case event during follow-up and 0 otherwise. The full likelihood is the sum of the terms l_i over all individuals. In the full-cohort situation, if interest is focused solely on the exposure effects β then the gain in efficiency of estimation from using the fully parametric model is small unless the regression coefficient is relatively large, when in any case statistical efficiency may be less critical.

A modification of this that allows the use of fully parametric models in nested case-control studies is possible. It can be achieved by using a weighted version of the full likelihood. First note that the probability that an individual i will ever be sampled to the nested case-control study, with random sampling of $m - 1$ controls within risk sets is

$$p_i = \begin{cases} 1, & Y_i = 1, \\ 1 - \prod_{t_j < \tilde{t}_i} (1 - (m-1)/\{n(t_j) - 1\}), & Y_i = 0, \end{cases} \qquad (7.18)$$

where for $Y_i = 0$ the product is over all event times at which the individual is at risk. The probability for non-cases is one minus the probability that an individual is not a case at each event time at which they are at risk. The above assumes, as is typical, that all cases in the underlying cohort are sampled to the nested case-control study. Using a parametric hazards model, as given above for the full-cohort study, a nested case-control study can be analysed by weighting each individual by the inverse of their probability p_i, giving the weighted pseudo-likelihood

$$\sum_{i \in \mathcal{N}} l_i / p_i, \qquad (7.19)$$

where $\mathcal{N}$ denotes the set of all individuals in the nested case-control study.

Inverse weighting by a sampling probability is often used in survey sampling settings. A brief review of inverse weighting in likelihood analyses is given in appendix Section A.7.

7.8.4 Aalen's linear hazard model

Another model of interest is Aalen's linear hazard model, in which the hazard function at time t for an individual with covariate set $\{x_1(t), \ldots, x_p(t)\}$ is

$$h\{t, x(t)\} = \alpha_0(t) + \alpha_1(t)x_1(t) + \cdots + \alpha_p(t)x_p(t), \qquad (7.20)$$

where $\alpha_0(t)$ is the hazard rate for an individual with baseline covariates, $\alpha_1(t)$ is the excess hazard at time t for an individual with covariate value $x_1(t)$ and so on. This model results in estimates of *excess* risk as opposed to *relative* risk. The cumulative excess rate, or excess risk, at time t due to a unit step change in $x_k(t)$ is

$$A_k(t) = \int_0^t \alpha_k(u) \, du, \quad k = 1, \ldots, p. \qquad (7.21)$$

Consider first a full-cohort study. We let $X(t)$ be a $(p + 1) \times N$ matrix with rows $(1 \quad x_{1i}(t) \quad \ldots \quad x_{pi}(t))$, $i = 1, \ldots, N$, where N is the total number of individuals in the cohort. It can be shown that an estimate for $A(t) = (A_0(t)A_1(t) \cdots A_K(t))^T$ is

$$\hat{A}(t) = \sum_{t_j \leq t} X^*(t_j) I_j, \qquad (7.22)$$

where I_j is a vector of indicators of event status at time t_j, taking the value 1 for the case at t_j and 0 elsewhere, and $X^*(t)$ is a generalized inverse of

$X(t)$, one option being $X^*(t) = \{X(t)^T X(t)\}^{-1} X(t)$. A simple modification to the above estimate can be made for use with nested case-control data. For this, each element of $X(t)$ is multiplied by $n(t)/m$, where $n(t)$ is the total number of individuals in the underlying population at risk at time t and m is the size of the sampled risk sets. We may then estimate $\hat{A}(t)$ as in (7.22).

Not all models for the occurrence of a point event, such as an individual's becoming a case, are directly hazards based, an accelerated failure-time formulation being one important alternative.

7.9 Extended use of controls

7.9.1 Preliminaries

In the nested case-control sampling designs considered so far, each case has its own set of controls chosen from the risk set available at the time in question. In the traditional conditional analysis based on the proportional hazards model, outlined in Section 7.3, each case is compared with its own controls. In this section we consider two ways in which an extended use of selected controls could be made:

(1) Controls selected at one time could equally serve as controls at all earlier times and at some later times where they remain in the risk set, thus increasing the number of controls that can be compared with each case in the conditional analysis. We refer to this as the forwards and backwards use of controls.

(2) A quite different situation is one in which we may wish to rc-usc controls selected for a cases in a study of one particular outcome as controls in the study of a second outcome within the same cohort.

7.9.2 Forwards and backwards use of controls

The standard analysis of data arising from a nested case-control study uses the partial likelihood (7.2). This ensures the nesting of the conditioning sets and thereby achieves zero correlation between the score contributions at distinct time points. Also, it accommodates time-varying exposures at least to the extent that the hazard of becoming a case may be assumed to depend solely on the current, appropriately defined, measure of exposure. If, however, the exposures of interest are not time dependent but constant for each individual then it is tempting to use controls chosen at one time also as controls for cases that occurred at earlier times, subject, of course, to any requirements of matching. Furthermore, controls could be used at subsequent times if they remained in the risk set. Such an extended use of controls, however, would conflict with the nesting of the conditioning sets, and hence in general distinct contributions to the score function would no longer be uncorrelated and special methods of analysis would be required. A fuller discussion is given in the appendix.

A further, possibly more critical, aspect of the above scheme is that unobserved time-dependent confounders that implicitly largely allowed for when cases and controls are matched in time might have a serious distorting effect if cases and controls are observed at an appreciable time apart. In some situations it may be possible, however, to obtain exposure and covariate information at a number of time points for individuals selected as controls.

7.9.3 Using controls for other outcomes

Prospective cohort studies in epidemiology usually follow individuals for a range of health outcomes and diagnoses, and separate nested case-control studies may be used to study risk factors for each separate outcome. This could be a poor use of resources, and it could be desirable to re-use some individuals selected in a study of one outcome in studies of different outcomes. The difficulty in re-using individuals sampled in a nested case-control study for a particular outcome is that controls are attached to individual cases of that original outcome at a specific event time, and so to use controls that have been decoupled from their original case requires a special analysis.

We outline one way of re-using individuals from a study of an outcome A in a study of a new outcome B and discuss how the resulting data can be analysed. We consider specifically the situation in which there exist data from a nested case-control study of outcome A, in which controls were selected for each case using random sampling from the risk set at each event time. The event times for outcome A are denoted t_{Aj} and for outcome B are denoted t_{Bj}. Suppose that no new controls are selected for cases of outcome B. We let i_j be the index of the jth case of type B and X a vector of exposures of interest for outcome B with corresponding log hazard ratio vector β. The cases and controls from the study of outcome A can be used in a weighted partial likelihood for a study of cases of outcome B, which is given by

$$\prod_j \frac{p_{i_j}^{-1} \exp\{\beta^T x_{i_j}(t_{Bj})\}}{\sum_{k \in \mathcal{N}_{ABj}} p_k^{-1} \exp\{\beta^T x_k(t_{Bj})\}}, \tag{7.23}$$

where $\mathcal{N}_{ABj}$ denotes the set of cases of type B at t_{Bj} plus all individuals from the nested case-control study for outcome A who were at risk at time t_{Bj} (that is, $t_{Ak} < t_{Bj}$). Individuals are weighted to accommodate the sampling procedure; one suggestion has been that the weighting should depend on their probability of being included in the sample. Under this scheme all cases of type B or type A receive a weight of 1, that is, $p_k = 1$, and all controls selected for outcome A are weighted by the inverse of their probability of ever having been sampled, using the weights in (7.18). The sandwich formula (see equation (A.20) in the appendix) should be used to estimate variances under this scheme.

The scheme can be extended to a situation in which controls have been selected for both outcomes using random-risk-set sampling and in which both sets of cases and controls are used in the analysis of each outcome. Extensions to more than two types of outcome are also possible.

The case-subcohort design, which is considered in Chapter 8, addresses the issue described in this subsection because the set of potential controls under that design is shared across cases of all types and therefore can be used for different outcomes.

7.10 Estimating absolute risk

In a nested case-control study, interest is typically in the relative effects of exposures. However, it is sometimes desirable to estimate the absolute risk of an outcome. We first outline this for a full-cohort study. The cumulative hazard at time t is

$$H(t, x(t)) = \int_0^t h_0(u) \exp\{\beta^T x(u)\}\, du.$$

To estimate $H(t, x(t))$ we require an estimate of the baseline hazard $h_0(t)$, which is not needed in a partial-likelihood analysis. To estimate $h_0(t)$ one possibility is a parametric analysis, using, for example, an exponential or Weibull form for the underlying distribution. Instead, we will replace $h_0(t)$ by a series of atoms of probability at the event times $t_1, \ldots, t_n$. An essentially equivalent alternative is to assume $h_0(t)$ to be constant between occurrence points. Let x_j^* denote as a reference level the vector of 'baseline' values for the covariates at t_j, for example the mean exposure value in the relevant risk set. Write the baseline hazard at time t_j as $h_j e^{-\beta^T x_j^*}$.

The full likelihood can be written as the product of the contributions from each failure time and is given by

$$\prod_{t_j} \prod_{k \in C_{1j}} h_j e^{-\beta^T x_j^*} e^{\beta^T x_k(t_j)} \prod_{k \in C_{0j}} \left(1 - h_j e^{-\beta^T x_j^*} e^{\beta^T x_k(t_j)}\right), \qquad (7.24)$$

where C_{1j} denotes the set of failures at t_j and C_{0j} denotes the set of non-failures, so that $C_{1j} \cup C_{0j} = R_j$. Here we allow for the possibility that more than one case may occur at the same time, in particular to allow for a grouping of time points.

The maximum likelihood estimate for h_j is given by the solution to

$$\frac{1}{\hat{h}_j} - \sum_{C_{0j}} \frac{e^{\beta^T \{x_k(t_j) - x_j^*\}}}{1 - \hat{h}_j e^{\beta^T \{x_k(t_j) - x_j^*\}}} = 0, \qquad (7.25)$$

assuming just one case at each time t_j. A first-order approximation gives

$$\hat{h}_j = \frac{1}{\sum_{k \in R_j} \exp\{\beta^T (x_k(t_j) - x_j^*)\}}. \qquad (7.26)$$

It follows that in a full-cohort study the cumulative hazard can be estimated as

$$\hat{H}\{t, x(t)\} = \sum_{t_j \leq t} \frac{\exp\{\beta^T (x(t) - x_j^*)\}}{\sum_{k \in R_j} \exp\{\hat{\beta}(x_k(t) - x_j^*)\}}. \qquad (7.27)$$

In a nested case-control study the link between the nested case-control sample and the underlying cohort is known, because the number of individuals at risk at each failure time is known. Therefore, although complete covariate information is available only for individuals in the nested case-control sample the fact that each sampled risk set is connected to the underlying cohort via the control sampling scheme provides information that allows estimation of the baseline hazard. The contribution to the baseline hazard estimate from individual k in the sampled risk set $\tilde{R}_j$ at time t_j is weighted by conditioning on the event that the sampled risk set comprises the observed set of individuals; this gives the conditional hazard function

$$P_j(k) h_j e^{-\beta^T x_j^*} e^{\beta^T x_k(t_j)}, \qquad (7.28)$$

where

$$P_j(k) = n(t_j)^{-1} \frac{p(\tilde{R}_j; k, \tilde{\mathcal{H}}_j)}{\sum_{l \in \tilde{R}_j} p(\tilde{R}_j; l, \tilde{\mathcal{H}}_j)}. \qquad (7.29)$$

Assuming that there are no tied event times, it follows that an estimate of the cumulative hazard is given by

$$\hat{H}(t, x(t)) = \sum_{t_j \leq t} \frac{\exp\{\beta^T (x(t) - x_j^*)\}}{\sum_{k \in \tilde{R}_j} \exp\{\beta^T (x_k(t) - x_j^*)\} P_j(k)}. \qquad (7.30)$$

The weights $P_j(k)$ depend on the nested case-control sampling scheme. Using standard nested case-control sampling, that is, random sampling without replacement from the risk set excluding the case at each failure time, $P_j(k) = n(t_j)/m$ where m is the size of the sampled risk set at each event time, including the case, and $n(t_j)$ is the size of the risk set. Under counter-matching $P_j(k) = n_{l_k}(t_j)/m_{l_k}$, where l_k denotes the sampling stratum for individual k.

7.11 Using full-cohort information

The traditional analysis of a nested case-control study, using (7.3), uses data only on sampled individuals. This ignores the potentially large amount of information that may be available on the full set of individuals in the underlying cohort. While typically a main exposure will be observed only in the nested case-control sample, other covariates such as sex, age, height, weight and smoking status may be observed for everyone in the underlying cohort. Further, in some situations a correlate of the expensive exposure may be available in the full cohort.

One way in which full-cohort information can be incorporated into the analysis of a nested case-control study is to use *multiple imputation* to impute any exposures that are missing in the full cohort. Multiple imputation is now widely used to deal with missing data. The key idea is that missing exposure measurements are imputed by drawing a random value from the distribution of the exposure that is conditional on all observed values, including the outcome. The imputation model is estimated in the fully observed data, here the nested case-control sample. To account for the uncertainty in the imputed values, a number M ($M > 1$) of imputed values are obtained for each missing data point, creating M imputed data sets for the full cohort. A full-cohort analysis using the usual partial likelihood is then performed within each imputed full-cohort data set, and the resulting hazard ratio estimates are combined. If $\hat{\beta}^{(k)}$ denotes the log odds ratio estimate from the kth imputed data set, $k = 1, \ldots, M$, then the pooled parameter estimate and its standard error are given by Rubin's rules,

$$\hat{\beta} = \frac{1}{M} \sum_{k=1}^{M} \hat{\beta}^{(k)}, \quad \mathrm{var}(\hat{\beta}) = A + \left(1 + \frac{1}{M}\right) B, \qquad (7.31)$$

where A and B represent respectively the within-imputation and between-imputation contributions to the variability. We have $A = \sum A_k/M$, where

A_k is the estimated variance of $\hat{\beta}^{(k)}$, and $B = \sum(\hat{\beta}^{(k)} - \hat{\beta})^2/(M - 1)$. The number of imputed data sets created is typically about 10.

We will focus on a single exposure X that is observed only in the nested case-control sample. Adjustment variables W are assumed to be available in the full underlying cohort. As noted above, the imputation model involves all observed variables including the outcome. Here the outcome for each individual in the cohort involves a time component. We let Y take the value 1 if an individual becomes a case during follow-up and 0 otherwise and t denote the time to event or censoring, whichever happens first. We do not describe the details of possible imputation models here but outline just one possibility, namely a model in which

$$X = \theta_0 + \theta_W W + \theta_Y Y + \theta_T H_0(t) + \theta_{WT} W H_0(t) + \epsilon \qquad (7.32)$$

where the θ_i are parameters and $H_0(t)$ is the cumulative baseline hazard at time t, which, as shown in Section 7.10, can be estimated from a nested case-control study. For a given individual, an imputed value $X^{(k)}$ is obtained as

$$\hat{\theta}_0^{(k)} + \hat{\theta}_W^{(k)} W + \hat{\theta}_Y^{(k)} Y + \hat{\theta}_T^{(k)} H_0(t) + \hat{\theta}_{WT}^{(k)} W H_0(t) + \epsilon^{(k)} \qquad (7.33)$$

where $\epsilon^{(k)}$ is a random draw from a normal distribution with mean 0 and variance $(\sigma_\epsilon^{(k)})^2$, and $(\sigma_\epsilon^{(k)})^2, \hat{\theta}_0^{(k)}, \hat{\theta}_W^{(k)}, \hat{\theta}_Y^{(k)}, \hat{\theta}_T^{(k)}, \hat{\theta}_{WT}^{(k)}$ are draws from the posterior distributions of the parameter estimates from the imputation model.

Multiple imputation using full-cohort information at the analysis stage of nested case-control studies has been shown to result in potentially substantial gains in efficiency, especially when the number of controls per case is small and when good correlates of the missing exposures are available in the full cohort. More complex imputation models appear to be needed when there are interaction terms or non-linear terms involving the partially missing exposure in the analysis model.

Example 7.5: Application of multiple imputation We will illustrate the multiple imputation approach using a nested case-control study of fibre intake and colorectal cancer within the EPIC-Norfolk cohort; this was first introduced in Example 3.1. This example was previously outlined by Keogh and White (2013). EPIC-Norfolk is a cohort of 25 639 individuals recruited during 1993–7 from the population of individuals aged 45–75 in Norfolk, UK, who have been followed up for a range of disease outcomes (see Day *et al.* 1999). Participants provided information about their dietary intake in two main ways: using a food frequency questionnaire (FFQ) and using a seven-day diet diary. Diet diaries are often considered to give better measurements of some aspects of dietary intake than FFQs, but they are comparatively very expensive to process. For this reason, some associations within EPIC-Norfolk, and within other cohorts, have been studied using diet diary measurements within nested case-control samples.

As described in the previous examples, 318 cases of colorectal cancer were registered up to the end of 2006 and four controls per case were sampled from the full cohort. Diet diary measurements of average daily food and nutrient intakes were obtained for individuals in the nested case-control sample. We will focus on the association between a 6 grams per day increase in average daily fibre intake (X) and colorectal cancer risk, with adjustment for variables W that include weight, height, smoking status, social class and level of education, plus four other dietary exposures. The non-dietary covariates were observed in the full cohort. In the full cohort FFQ measurements of the dietary exposures were also observed, providing surrogate measurements that we included in the imputation model as part of W.

An imputation model similar to that in (7.32) was used to impute the missing exposure in the full cohort. Further details are given in Keogh and White (2013). The standard nested case-control analysis resulted in a log hazard ratio estimate of -0.173 with standard error 0.104 and 95% confidence interval $(-0.377, 0.030)$. Using the multiple imputation approach gave a hazard ratio estimate -0.167 with standard error 0.072 and 95% confidence interval $(-0.308, -0.026)$. Thus, using the multiple imputation analysis there is a reduction in the standard error of the estimate relative to the traditional nested case-control analysis. Many of the adjustment variables were observed for nearly everyone on the full cohort, and some of these variables must be highly correlated with the main exposure of interest.

Notes

Section 7.2. The nested case-control design appears to have been first considered by Liddell *et al.* (1977) in a discussion of methods of analysis for longitudinal studies from a prospective or retrospective standpoint, with an application in a study of occupational exposure to asbestos and lung cancer risk as given in Example 7.1.

Sections 7.3–7.5. The proportional hazards model and partial likelihood method are due to Cox (1972, 1975). The asymptotic properties of the partial likelihood were discussed by Cox and Oakes (1984). There is a large specialized literature associated with survival analysis and, more broadly, event-history analysis. See Oakes (1981). The method of analysis for nested case-control studies using partial likelihoods was described explicitly in an addendum to the paper of Liddell *et al.* (1977) by Thomas (1977). It was formalized by Prentice and Breslow (1978). The asymptotic properties of the parameter estimates arising from a partial-likelihood analysis of a nested case-control study were set out by Goldstein and Langholz (1992). See also Borgan *et al.* (1995) and Langholz and Goldstein (1996). Much of the statistical literature on the nested case-control design formulates the problem using counting process notation, which was set out by Borgan *et al.* (1995) and is used throughout the papers of Borgan, Langholz and Goldstein. See also the book on counting processes by Andersen *et al.* (1993). Prentice (1986b) proposed an alternative to the random sampling designs discussed in this section. Prentice noted that in small studies there may be some correlation between successive score terms in the partial likelihood for a nested case-control study, and he considered sampling designs which would have the result that the complete set of sampled risk sets are disjoint. To achieve such a design, at a given failure time any individual who had been chosen for the sampled risk set at an earlier time would not be eligible for selection. Additionally, to ensure disjointness of the sampled risk sets, any case previously sampled as a control would not be eligible to be included in the study as a case at its failure time. Prentice considered this second condition to be wasteful and decided to drop it from his sampling scheme. The result is that the likelihood under this scheme is of the form (7.3) but does not have a partial-likelihood interpretation, that is, it is a pseudo-partial likelihood. Another possibility, in theory at least, is to sample randomly with replacement from the full risk set R_j. Here, the case could potentially also serve as one of its own controls. Then, if s_i denotes the number of appearances of individual i, the selection probability is

$$\left(\frac{1}{n_j}\right)^{m-1} \frac{(m-1)!}{s_1! \cdots s_{n_j}!}$$

and it follows that the design probabilities in (7.2) cancel out, giving (7.3). See Robins *et al.* (1986).

Some alternative sampling designs for use in open cohorts were considered by Robins *et al.* (1989).

Section 7.6. Counter-matching in a nested case-control study was first described by Langholz and Borgan (1995) and Langholz and Clayton (1994). Langholz and Borgan (1995) suggested a potential strategy for selecting cut points for a continuous exposure to be used in counter-matched sampling. Steenland (1997) gave an example of counter-matching for a continuous exposure using a surrogate of exposure, in which two methods of choosing the cut points of the exposure to create sampling strata were considered: (i) dividing the surrogate exposure into quartiles within each risk set and performing counter-matching within quartiles, or (ii) dividing the surrogate exposures for the cases into quartiles and applying these cut points across all risk sets. The second method was found to give better precision. Langholz and Goldstein (1996) and Borgan and Olsen (1999) investigated the efficiency of the counter-matching design relative to standard nested case-control sampling in the two situations considered in this section. The efficiency of counter-matching versus standard nested case-control sampling for interaction terms was assessed by Cologne and Langholz (2003), Cologne *et al.* (2004) and Andrieu *et al.* (2001) in the context of gene-environment interactions.

Quota sampling was first described by Borgan *et al.* (1995). A framework for the analysis of case-control studies under a number of designs, including counter-matching and quota sampling, is described in a very useful paper by Langholz and Goldstein (2001).

Other more elaborate sampling procedures include the use of a second stage of sampling within a nested case-control study, within which additional information is obtained for an extended study. See Langholz and Goldstein (1996). There is a connection here with Section 6.5.

Section 7.7. Lubin and Gail (1984) derived the bias in hazard ratio estimates that can arise if the restrictions discussed in this section are placed on the eligibility of individuals for control selection at a given failure time. See also Robins *et al.* (1986) for further discussion.

Section 7.8.2. Aranda-Ordaz (1983) gave a method for the choice between alternative hazard-based formulations. Cox and Wong (2004) discussed the stability of ratios of regression coefficients. Muirhead and Darby (1987) compared additive and proportional hazards models for studies of radiation-induced cancer.

Section 7.8.3. The use of a weighted pseudo-likelihood to enable the use of fully parametric models in nested case-control studies was suggested by Samuelsen (1997), who also gave the probabilities (7.18) and provided asymptotic results and simulation studies of its efficiency relative to the standard analysis. See Kalbfleisch and Lawless (1988) for a more general discussion of the use of weighted pseudo-likelihoods in studies of disease incidence, with a focus on studies with a retrospective element. Samuelsen (1997) also suggested that the weights derived in his paper can also be applied in a partial-likelihood analysis of the nested case-control data, assuming the proportional hazards model; the resulting pseudo partial likelihood is

$$\prod_j \frac{\exp\{\beta^T x_{i_j}(t_j)\}}{\sum_{k \in \tilde{R}_j} p_k^{-1} \exp\{\beta^T x_k(t_j)\}}. \tag{7.34}$$

This pseudo-partial likelihood has been found to offer some potential gain in efficiency relative to the unweighted partial likelihood analysis. This is so presumably because information about the underlying event rates is contained in the inclusion probabilities p_k.

Section 7.8.4. Aalen's linear model (1989), was extended to the context of nested case-control studies by Borgan and Langholz (1997). See also Zhang and Borgan (1999). Solomon (1984) showed the stability of ratios of regression coefficients relative to proportional hazards and accelerated life models.

Section 7.9.2. Langholz and Thomas (1991) investigated two ways of analysing nested case-control study data, which, it was thought, could improve efficiency. The first method ('retained nested case-control sampling') augments the sampled risk sets by adding to them all individuals sampled at earlier failure times. While this design results in a likelihood with a partial-likelihood interpretation, it was not found to result in a gain in efficiency relative to the standard nested case-control design using sampling without replacement from non-cases at each failure time. The second method ('augmented nested case-control sampling') re-uses sampled controls both forwards in time *and* backwards in time so that any individual sampled as a control at a given failure time is included in the augmented sampled risk set for any case for which that individual appears in the risk set. The second design was found to result in either a slight gain or slight loss in efficiency relative to the standard nested case-control design, depending on the exact circumstances of the prospective cohort.

Section 7.9.3. Methods for allowing the re-use of controls for different outcomes in nested case-control studies were considered by Saarela *et al.* (2008). See also Salim *et al.* (2009) and Støer and Samuelsen (2012). Reilly *et al.* (2005) gave a method for allowing the re-use of data for different outcomes in non-nested case-control studies.

Section 7.10. The argument leading to the baseline hazard estimator for a full-cohort study follows that given by Cox (1972). The estimator (7.26) was suggested by Breslow (1972). Estimation of the cumulative hazard in a nested case-control study was considered by Borgan and Langholz (1993) for nested case-control studies with random sampling from risk sets, and by Borgan *et al.* (1995) and Langholz and Borgan (1997) for general nested case-control sampling schemes. Borgan and Olsen (1999) showed that counter-matching can be less efficient in estimating the cumulative hazard compared with using random sampling from the risk sets.

Section 7.11. Multiple imputation was devised by Rubin (1987); thus the method for combining estimates from multiple imputed data sets in (7.31) is referred to as 'Rubin's rules'. It has been shown that multiple imputation results in unbiased pooled parameter estimates and, provided that the imputation model is reasonable, correct standard errors. A summary is given in White *et al.*

(2011), for example. Use of multiple imputation when the outcome involves a time component was outlined by White and Royston (2009). The use of multiple imputation in nested case-control studies was described by Keogh and White (2013) and was found to result in gains in efficiency relative to the standard analysis. Multiple imputation has also been extended to the case of non-monotone missing values in multiple exposures or covariates and software is available in commonly available statistical packages. For methods for analysing missing data in proportional hazards regression see, for example, Robins *et al.* (1994), Prentice (1982), Zhou and Pepe (1995) and Chen and Little (1999). Liu *et al.* (2010) and Scheike and Juul (2004) considered different methods, specifically for nested case-control studies.

8

Case-subcohort studies

- The case-subcohort design, often called simply the case-cohort design, is an alternative to the nested case-control design for case-control sampling within a cohort.
- The primary feature of a case-subcohort study is the 'subcohort', which is a random sample from the cohort and which serves as the set of potential controls for all cases. The study comprises the subcohort plus all additional cases, that is, those not in the subcohort.
- In an analysis using event times the cases are compared with members of the subcohort who are at risk at their event time, using a pseudo-partial likelihood. This results in estimates of hazard ratios.
- An advantage of this design is that the same subcohort can be used to study cases of different types.
- A simpler form of case-subcohort study disregards event times and is sometimes referred to as a case-base study or hybrid epidemiologic design. In this the subcohort enables estimation of risk ratios and odds ratios.

8.1 Preliminaries

In this chapter we continue the discussion of studies described broadly as involving case-control sampling within a cohort. In the nested case-control design, discussed in Chapter 7, cases are compared with controls sampled from the risk set at each event time. A feature of the nested case-control design is that the sampled controls are specific to a chosen outcome and therefore cannot easily be re-used in studies of other outcomes of interest if these occur at different time points; in principle, at least, a new set of controls must be sampled for each outcome studied though some methods have been developed that do enable the re-use of controls.

The case-subcohort design now to be discussed involves an alternative way of selecting a case-control-type sample within a cohort that avoids

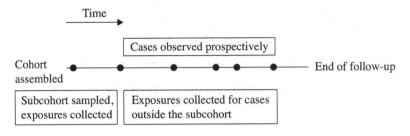

Figure 8.1 Essential outline for a case-subcohort study.

some limitations of a nested case-control study. As in the previous chapter, we consider a source population in which the occurrences of events of interest are observed. The source population for a case-subcohort study may be an assembled cohort under prospective follow-up for a range of outcomes or a well-defined population of individuals not formally assembled as participants in a study. These two settings were discussed in Section 7.1. The cases are those individuals who have the event of interest during the follow-up period. In this different form of study, a set S of individuals called the *subcohort* is sampled at random and without replacement from the cohort at the start of follow-up of the source population. Alternatively, the selection of the subcohort may occur after some or all follow-up has taken place but still without regard to any information acquired during follow-up, such as case status, and thus as if it had been done at the start of follow-up. Because the subcohort is a random sample from the cohort it will typically contain some cases of the outcome of interest. A case-subcohort study thus consists of the subcohort plus all additional cases occurring in the full cohort but outside the subcohort. See Figure 8.1.

The subcohort is the defining feature of a case-subcohort study, and it serves as the set of potential controls for all cases that are observed in the full cohort. Information on exposures and covariates is obtained for all cases and members of the subcohort. Sometimes time-varying exposure or covariate information may be collected, and we consider this further below. The subcohort does not receive any special treatment, in the sense that individuals are in principle not aware that they are in the subcohort nor typically are they are monitored more closely than individuals outside the subcohort.

How the comparison between cases and controls is achieved in a case-subcohort study, and how 'time' is handled, is discussed in more detail in the next section.

8.2 Essential outline of the case-subcohort design

We consider two approaches to analysing data from a case-subcohort study, depending on whether the times at which cases occur are available and if so whether they are to be used. In the first approach event times for cases are disregarded and in the second the times at which cases occur are indeed used.

(1) *Disregarding event times.* We suppose first that the times at which cases are identified are not used (or perhaps even not known). That is, it is assumed that a cohort of individuals is followed for a fixed time period and that in principle at some point all subcohort individuals are checked for case-control status, explanatory variables recorded and the corresponding data collected on all cases in the full cohort. This type of study has also been referred to as a *case-base study*, an *inclusive case-control design* or a *hybrid epidemiological design*.

(2) *Using event times.* In a more elaborate, and in principle more informative, situation the times at which cases arise are used. Then, at a given event time t_j the control set for the observed case i_j is the set S_j of individuals who are in the subcohort and who are at risk at that time, including individuals who later become cases. The comparison set for the case i_j at time t_j is thus $S_j \bigcup i_j$.

Approach (1), in which event times are disregarded, is in a sense similar to a standard case-control design except that controls are sampled at the start of some notional follow-up period rather than at the end. Approach (2), in which event times are used, has more in common with a *nested* case-control study. This second study is what is most commonly referred to a 'case-cohort study'. There is an intermediate situation in which information on how long individuals in the subcohort are at risk is used in the analysis, but actual event times are not used; this allows a simple analysis that deals to some extent with 'time', under some strong assumptions. An important benefit of using a case-subcohort study incorporating event times is that the analysis can incorporate individuals joining and leaving the population, as well as time-varying explanatory variables.

The two types of case-subcohort study described above and the intermediate situation allow the estimation of different quantities. Like a standard case-control study, a case-subcohort study that disregards time enables the estimation of an odds ratio. However, it also permits estimation of the *risk ratio*, sometimes referred to as the relative risk. A case-subcohort study using event times permits estimation of the *rate* or *hazard ratio*. The intermediate case enables the estimation of rate ratios under some assumptions.

Example 8.1: Case-cohort study using event times   Armstrong *et al.* (1994) used a case-subcohort study using time-to-event data to study whether aluminium plant workers exposed to coal tar pitch volatiles were at increased risk of death from lung cancer. The source population was over 16 000 men who had worked for at least one year in an aluminium production plant between 1950 and 1979 in Quebec, Canada. There were 338 lung cancer deaths observed during follow-up of this population between 1950 and 1988. The subcohort was a random sample of 1138 men from that population.

Two indices of exposure to coal tar pitch volatiles were studied. Company experts estimated the exposure indices for different job types and across calendar time. These estimates were combined with work histories to estimate cumulative exposures. It was of particular interest in this study to investigate confounding by smoking, and so smoking histories were obtained from company medical records.

The time scale in this study was age, and each case was compared with members of the subcohort still alive at the case's age. This required estimation of the cumulative exposures for all at-risk subcohort members at each case's age at death.

8.3 Analysis disregarding time

Figure 8.2 illustrates a case-subcohort study in which event times were disregarded. We will focus on a time period, preferably short, during which cases are observed. Individuals who do not become cases are here assumed to be observed for the full time period of interest. In other words, here we do not allow for a source population in which individuals leave the population before the end of the follow-up time, that is, are censored, or join the population after the time origin. We consider for simplicity a binary explanatory variable X and denote cases by $Y = 1$ and non-cases, by $Y = 0$. Suppose that in the population we have

$$\Pr(X = x, Y = y) = \pi_{xy}. \tag{8.1}$$

We denote the number of individuals in the full cohort by n and the number in the subcohort by m. Then individuals are classified into cells as follows: four cells in the subcohort corresponding to the possible pairs of values of X and Y; two cells outside the subcohort but with $Y = 1$ and X observed;

Table 8.1 *Data from a case-subcohort study, disregarding event times.*

	Subcohort		Outside subcohort	
	Controls	Cases	Controls	Cases
Unexposed	$m_0 - d_{0S}$	d_{0S}	Unobserved	d_{0C}
Exposed	$m_1 - d_{1S}$	d_{1S}	Unobserved	d_{1C}

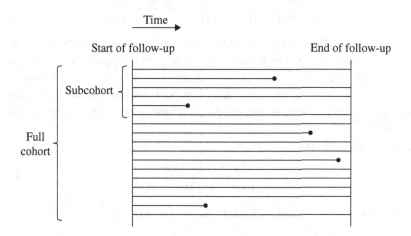

Figure 8.2 A case-subcohort study within a population in which cases (•) occur over the course of follow-up and all individuals are observed for the entire follow-up period. The lines represent the time period over which individuals are observed.

one cell outside the subcohort with X unobserved and with $Y = 0$. Suppose that in the subcohort there are d_{xS} cases with $X = x$ and $m_x - d_{xS}$ controls with $X = x$ and that among the cases outside the subcohort there are d_{xC} with $X = x$. The data are summarized in Table 8.1.

Assuming the mutual independence of individuals, the full likelihood of the data summarized in Table 8.1 has the multinomial form

$$\pi_{00}^{m_0-d_{0S}} \pi_{10}^{m_1-d_{1S}} \pi_{01}^{d_{0S}} \pi_{11}^{d_{1S}} \pi_{01}^{d_{0C}} \pi_{11}^{d_{1C}} (\pi_{00} + \pi_{10})^{n-m.-d.c}, \tag{8.2}$$

where the first four factors refer to the subcohort and the last three to the remainder of the cohort; $m. = m_0 + m_1, d.c = d_{0C} + d_{1C}$.

Two quantities can be estimated using the data in Table 8.1, an odds ratio and a risk ratio. The odds ratio for a binary exposure has been discussed in detail in previous chapters and is given by

$$e^\psi = \frac{\pi_{00}\pi_{11}}{\pi_{01}\pi_{10}}. \tag{8.3}$$

The risk ratio is defined as

$$\phi = \frac{\Pr(Y = 1 | X = 1)}{\Pr(Y = 1 | X = 0)} \tag{8.4}$$

and can be expressed alternatively as

$$\phi = \frac{\Pr(X = 1 | Y = 1)\Pr(X = 0)}{\Pr(X = 0 | Y = 1)\Pr(X = 1)} = \frac{\pi_{11}\pi_{0\cdot}}{\pi_{01}\pi_{1\cdot}}. \tag{8.5}$$

The subcohort, being a random sample from the population, provides information on the marginal probability of exposure in the population, enabling estimation of ϕ. This is in contrast with the case-control studies discussed in earlier chapters, where the connection between the case-control study and the underlying cohort was unknown, allowing estimation only of odds ratios.

The odds ratio is estimated as

$$e^{\hat\psi} = \frac{(d_{1C} + d_{1S})(m_0 - d_{0S})}{(d_{0C} + d_{0S})(m_1 - d_{1S})}. \tag{8.6}$$

Using the results of earlier chapters, the variance of $\log \hat\psi$ is

$$\frac{1}{d_{1C} + d_{1S}} + \frac{1}{m_0 - d_{0S}} + \frac{1}{d_{0C} + d_{0S}} + \frac{1}{m_1 - d_{1S}}. \tag{8.7}$$

The risk ratio can be estimated by

$$\hat\phi = \frac{(d_{1C} + d_{1S})m_0}{(d_{0C} + d_{0S})m_1}. \tag{8.8}$$

The two quantities $e^{\hat\psi}$ and $\hat\phi$ will be very similar if cases are rare. If cases are common then the risk ratio will typically be of more interest than the odds ratio, which does not have such an easy interpretation in this context.

The cases appearing in the subcohort contribute to both the numerator and the denominator of the risk ratio estimate. Hence, the variance of the log risk ratio is not given by the familiar formula. It can be shown that it is in fact

$$\frac{1}{d_{1C} + d_{1S}} + \frac{1}{d_{0C} + d_{0S}} + \left(\frac{1}{m_1} + \frac{1}{m_0}\right)\left(1 - 2\frac{d_{\cdot S}}{d_{\cdot S} + d_{\cdot C}}\right), \tag{8.9}$$

where $d_{\cdot S} = d_{0S} + d_{1S}$.

Note that an alternative estimator for the risk ratio uses only cases outside the subcohort to estimate $\Pr(X = x | Y = 1)$, since these individuals are in fact a random sample of the cases in the full cohort. The alternative estimator is therefore

$$\tilde{\phi} = \frac{d_{1C} m_0}{d_{0C} m_1}. \tag{8.10}$$

The variance of $\log \tilde{\phi}$ can be estimated using the standard formula for odds ratios, and is

$$\frac{1}{d_{1C}} + \frac{1}{m_0} + \frac{1}{d_{0C}} + \frac{1}{m_1}. \tag{8.11}$$

Both risk ratios, as given in (8.8) and (8.10), can be viewed in a sense as odds ratios; this simply requires us to define carefully the case and control groups. In the first version, (8.8) is an odds ratio where the case group is all cases observed during the follow-up time and the 'control' group is the subcohort, though this may in fact contain some cases. In the second version, (8.10) is an odds ratio where the case group is only those cases outside the subcohort, and the control group is the subcohort.

Note that the size n of the source population is not required for estimation of either the odds ratio or the risk ratio in the above setting. Hence this approach can be used within any population, whether or not it is fully enumerated, provided that a random sample can be obtained. If event times are not going to be involved in the analysis, a case-subcohort study could be used in a more totally retrospective context, where both the subcohort and the additional cases are identified retrospectively.

It will usually be necessary to consider an adjustment for other variables in estimating either the odds ratio or risk ratio. In some simple situations estimates can be found within strata defined by adjustment variables, and the estimates combined if appropriate. Methods for combination of odds ratios across strata were discussed in Chapter 2. In most realistic settings a regression approach will be required, which allows for continuous exposures and adjustment for multiple covariates. For the estimation of odds ratios, the main results of Chapter 4 can be applied. If the case and control groups are defined in such a way that they are overlapping, such as when the risk ratio in (8.8) is viewed as an odds ratio, sandwich variance estimators (see (A.20)) should be used to take into account that some individuals appear in the analysis as both cases and controls.

For the analyses described above we assumed a population that is unchanging over the observation period, as illustrated in Figure 8.2. However, unless the period of observation is short we will usually be faced with a changing population. If the underlying population is an assembled cohort observed over a long period of time then individuals will leave the population due to censoring. In some situations the population will have an influx of individuals after the start of follow-up. The above analysis does not easily accommodate these more complex situations.

One possibility is to sample individuals to the subcohort in such a way that their probability of appearing in the subcohort is proportional to their total time at risk. This possibility was in fact considered in Section 2.10. The risk ratio estimated under these situations not using event times is interpretable as a rate ratio only if the hazard rate is constant over time or if it happens that the ratio of exposed to unexposed individuals at risk stays the same over time and the event rates within the exposed and unexposed remain the same over time. Neither this situation nor that described above allows for exposures that may change appreciably in an individual over the period of follow-up, nor does either permit the estimation of rate ratios without additional assumptions being made.

8.4 Case-subcohort study using event times

8.4.1 The main result

We now consider the more complicated analysis applying when each case is compared with controls that are appropriate at the time point defined by the case. This is the more usual form of case-subcohort analysis, and can handle time-dependent exposures and changing populations.

Figure 8.3 illustrates the formation of comparison sets for each case under a case-subcohort study using event times. In the illustration, all individuals in the cohort are present at the start of follow-up. However, influx into the population can be accommodated by allowing any individual joining the population after the start of follow-up to be sampled to the subcohort with the same probability as individuals who were present at the start of follow-up. If sampling to the subcohort is in fact performed retrospectively then this particular issue is averted.

At a case occurrence time t_j, S_j denotes the set of individuals in the subcohort who are at risk at time t_j. The case, indexed by i_j, may be an individual inside or outside the subcohort. We also write

$$\tilde{S}_j = \begin{cases} S_j & \text{if } i_j \in S, \\ S_j \bigcup i_j & \text{if } i_j \notin S. \end{cases}$$

We refer to $\tilde{S}_j$ as the *case-subcohort risk set* at time t_j.

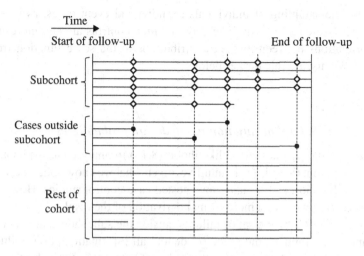

Figure 8.3 A case-subcohort study using event times. The horizontal lines show individual follow-up. The cases are indicated by • and the non-cases used in the comparison set at each failure time are indicated by ◇. Censoring occurs when a horizontal line stops before the end of follow-up. The broken vertical lines pass through the full risk set at each event time.

We focus here on use of the proportional hazards model for survival times in the analysis of case-subcohort data, though other models are possible. For an individual with a possibly time-dependent vector of exposures and covariates $x(t)$ at time t, the hazard function at time t is

$$h(t; x(t)) = h_0(t) \exp\{\beta^T x(t)\}. \tag{8.12}$$

Use of the proportional hazards model in the conditional analysis of prospective cohort studies and of nested case-control studies was discussed in Chapter 7. To analyse the data arising from a case-subcohort study we use the *pseudo-partial likelihood*

$$\tilde{\text{lik}}(\beta) = \prod_j \frac{\exp\{\beta^T x_{i_j}(t_j)\}}{\sum_{k \in \tilde{S}_j} \exp\{\beta^T x_k(t_j)\}}, \tag{8.13}$$

where the product is over all case occurrence times. The denominator sums over all individuals at risk in the subcohort plus the case itself if this appears outside the subcohort. In the special case in which all the individuals in the cohort are taken to be the subcohort, (8.13) becomes the usual partial likelihood and $\tilde{S}_j$ in the denominator could be replaced by the full risk set R_j. Note that in a case-subcohort study the comparison set for each case is determined from the sampling of the subcohort at the study origin rather than by the sampling of individuals at individual event times, as it is in a nested case-control study. The way in which controls are chosen results in correlations between distinct contributions to the score function from (8.13). We discuss this further below.

8.4.2 Formulation using design probabilities

The use of the pseudo-partial likelihood (8.13) to analyse case-subcohort data using event times is fairly intuitive, given what we know about the analysis of full cohort studies and, now, nested case-control studies. However, in this section we give a more formal derivation of the result.

First we let C_j be the set of failures outside the subcohort that occur up to time t_j. At failure time t_j let $\tilde{\mathcal{H}}_j$ denote all information about failures and censorings and explanatory variables of interest in the cohort up to just before time t_j, plus the information on whether the case *at* time t_j is in the subcohort. Some exposure variables will be observed only for those individuals in the subcohort S and for the set of extra cases C_j. Information about which information is in the subcohort is not contained within $\tilde{\mathcal{H}}_j$; otherwise, anyone falling outside the subcohort would automatically be identified as the case at a given event time conditional on $\tilde{\mathcal{H}}_j$.

We define a pseudo-partial likelihood by taking a contribution from the case-subcohort risk set $\tilde{S}_j$ at each time t_j consisting of the probability that, given that exactly one case occurs at the time t_j and given that the case-subcohort risk set is that observed, the case is in fact individual i_j and not one of the other individuals in $\tilde{S}_j$. We define $p(\tilde{S}_j; i_j, \tilde{\mathcal{H}}_j)$ as the probability of observing the particular case-subcohort risk set $\tilde{S}_j$ given the identity i_j of the case at t_j and given all the history summarized in $\tilde{\mathcal{H}}_j$. This gives the pseudo-partial likelihood contribution

$$\frac{p(\tilde{S}_j; i_j, \tilde{\mathcal{H}}_j) \exp\{\beta^T x_{i_j}(t_j)\}}{\Sigma_{k \in \tilde{S}_j} p(\tilde{S}_j; k, \tilde{\mathcal{H}}_j) \exp\{\beta^T x_k(t_j)\}}. \tag{8.14}$$

Under the assumption that individuals were selected to the subcohort using random sampling and that an individual's being in the subcohort has no effect on his or her survival time, the number of individuals at risk in the subcohort at a given time is given by a random sample of the individuals at risk in the full cohort at that time. Let $n(t_j)$ and $m(t_j)$ denote the number of individuals at risk in the cohort and in the subcohort respectively at failure time t_j. When the case at time t_j is in the subcohort we have $|\tilde{S}_j| = m(t_j)$

and the weight is

$$p(\tilde{S}_j; i_j, \tilde{\mathcal{H}}_j) = \binom{n(t_j)}{m(t_j)}^{-1}. \tag{8.15}$$

When the case i_j at time t_j is *not* in the subcohort, we have $|\tilde{S}_j| = m(t_j) + 1$. The only way for the sampled risk set $\tilde{S}_j$ still to be as observed had individual k from the subcohort instead been the case would be for individual i_j to have been sampled to the subcohort instead of individual k. It follows that the weights $p(\tilde{S}_j; k, \tilde{\mathcal{H}}_j)$ are as in (8.15). The result is that the terms $p(\cdot)$ in (8.14) all cancel, resulting in the unweighted pseudo-partial likelihood (8.13).

The information $\tilde{\mathcal{H}}_k$, referring to the failure time t_k, does not include information on whether the cases at earlier failure times t_j, $j < k$, were in the subcohort; for it to do so would give too much information about the identity of the case. It follows that $\tilde{\mathcal{H}}_j$ is not nested in $\tilde{\mathcal{H}}_k$, $j < k$. A consequence is that the individual pseudo-score function contributions $\tilde{U}_j(\beta)$, where $\tilde{U}_.(\beta) = \partial \log\{l\tilde{i}k(\beta)\}/\partial\beta = \sum_j U_j(\beta)$, are correlated and the sandwich formula (A.20) is required for variance estimation. We refer to the appendix for further details on pseudo-likelihood. The sandwich variance estimator is not straightforward to compute, at least by hand, and it does not easily extend to general weighted pseudo-partial likelihoods, where the weights $p(\cdot)$ in (8.14) do not cancel; these are discussed briefly in Section 8.6. In many ways a computationally simpler estimate of variance can be obtained by finding the influence of each individual on the estimate $\hat{\beta}$, that is, the effect of omitting the individual. The precision of the estimate of interest can be deduced from the magnitude of the influences. We omit the details here. It should be emphasized that statistical software packages now include functions specifically for the analysis of case-subcohort data, so that the complexity involved in variance calculations is no longer a drawback to their application.

Various relatively *ad hoc* modifications of the weights $p(\cdot)$ in (8.14) have been suggested with the objective of simplifying the analysis. These modifications are based on weighting individuals according to their probability of being sampled to the case-subcohort study, that is, on Horwitz–Thompson weighting, which is outlined briefly in appendix section A.7. One such approach uses a pseudo-partial likelihood of the form

$$\prod_j \frac{\exp\{\beta^T x_{i_j}(t_j)\}}{\exp\{\beta^T x_{i_j}(t_j)\} + (1/w)\sum_{k \in \tilde{S}_j \backslash i_j} \exp\{\beta^T x_k(t_j)\}}. \tag{8.16}$$

where w is the probability of an individual being included in the subcohort. That is, in the denominator the case itself has weight 1 and individuals in the subcohort have a weight that is their inverse probability of being included in the subcohort.

8.5 Size of the subcohort

We have not yet discussed the size of the subcohort relative to the full cohort in a case-subcohort study. The size of the subcohort will typically be restricted by the available resources. However, the subcohort should not be too small relative to the number of cases because this may have the result that some cases are without any individuals for comparison in the subcohort if there is other censoring or if the time scale is 'age'.

The asymptotic relative efficiency of a case-subcohort study compared with the equivalent full cohort study depends on a number of factors in addition to subcohort size, including the changing nature of the population, the event rate and the association between the exposure and the event rate.

We focus on the simple situation of a cohort in which there is no influx into the population and no censoring and with $\beta = 0$.

We may regard the comparison between an estimate from a subcohort study and a corresponding estimate from a full cohort study as being similar in its dependence on sample size to a comparison of two sample means of different fixed sizes. These are determined by π, the proportion of cases in the full cohort, and p, the proportion of the cohort chosen to form the subcohort. In the full cohort the samples compared are of sizes $N\pi$ and $N(1-\pi)$, say, leading to a variance of estimation proportional to $1/(N\pi) + 1/\{N(1-\pi)\}$. In the case-subcohort study the corresponding sample sizes are $N\pi$ and $pN(1-\pi)$ and the variance is proportional to $1/(N\pi) + 1/\{pN(1-\pi)\}$. The ratio of the resulting variances for a given p to that at $p = 1$ (the full cohort) is

$$1 + \pi(1/p - 1). \tag{8.17}$$

The asymptotic efficiency of the case-subcohort study relative to the full-cohort study is therefore the reciprocal of the above and is illustrated in Figure 8.4.

Example 8.2: Effect of subcohort size on estimated standard errors We will use data on the association between alcohol intake and breast cancer

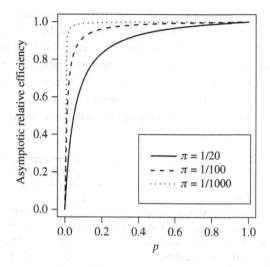

Figure 8.4 Approximate asymptotic efficiency of a
case-subcohort study relative to that of a full-cohort study for a
binary exposure when the hazard ratio is 1 ($\beta = 0$). The
proportion of individuals in the subcohort is p and the proportion
of individuals who become a case is π.

among women in the EPIC-Norfolk cohort to illustrate the effect of subcohort size on estimated standard errors. We also show the incorrect standard errors that would be obtained if the variance of the parameter estimates were estimated using the inverse of the information matrix.

As already mentioned, EPIC-Norfolk is a cohort of over 25 000 individuals recruited during the mid 1990s. Here we focus on the group of 12 433 female participants, within which 469 breast cancer cases were recorded up to the end of February 2010. Participants completed a food frequency questionnaire at recruitment to the study, on which information on alcohol intake was obtained. In reality, therefore, the main exposure is observed in the full population and a full cohort analysis (restricted to women) can be performed. For this illustration we took case-subcohort samples within the cohort (restricting to women only) using subcohorts of size 5%, 10%, 20% and 50% of the full cohort. For each subcohort size, 1000 subcohort samples were taken.

The analysis used the pseudo-partial likelihood in (8.13) and adjusted also for a number of potential confounders, which we do not list here. Table 8.2 shows the log hazard ratio estimate and its standard error under

Table 8.2 *Results from studies of alcohol intake (measured in units of 10 grams per day) and breast cancer in EPIC-Norfolk. Case-subcohort studies with subcohorts of different sizes were sampled from the cohort, with each sampling scenario repeated 1000 times. Results are mean log hazard ratio (HR) estimate and standard errors across the 1000 samples. We show estimates of correct standard errors using sandwich estimators (SE) and incorrect standard errors using the inverse information matrix (SE*).*

	log HR	SE	SE*
Full cohort analysis	0.124	NA	0.046
Case-subcohort analysis, subcohort size 5%	0.136	0.072	0.048
Case-subcohort analysis, subcohort size 10%	0.128	0.059	0.047
Case-subcohort analysis, subcohort size 20%	0.125	0.051	0.047
Case-subcohort analysis, subcohort size 50%	0.124	0.045	0.046

a full cohort analysis and also the average of the 1000 log hazard ratio estimates and the standard errors found under the 1000 generated case-subcohort studies.

In this example a subcohort consisting of 20% of the full cohort achieves results close to those from the full cohort analysis on average, and for 50% there is virtual identity. If incorrect variances are used, that is, if the case-subcohort data is analysed as a full cohort study, then the standard errors of estimates may be quite severely underestimated if the subcohort is fairly small relative to the full cohort.

8.6 Stratified case-subcohort studies

Extensions to the case-subcohort studies discussed previously include the use of stratification at the analysis stage, or the sampling stage or both.

A stratified analysis of case-subcohort data is possible in which each case is compared only with individuals in the same stratum of the subcohort as the case. No major adjustment to the methods described previously is needed; in particular no special weighting of the pseudo-partial likelihood is required. The pseudo-partial likelihood in this case is

$$\prod_j \frac{\exp\{\beta^T x_{i_j}(t_j)\}}{\sum_{k \in \tilde{S}_{s_j}} \exp\{\beta^T x_k(t_j)\}}, \qquad (8.18)$$

where $\tilde{S}_{s_j}$ denotes the subset of the case-subcohort risk set at time t_j which is in the same stratum as the case, denoted by s_j. This type of analysis is broadly analogous to matching on additional covariates when controls are sampled at each event time in a nested case-control study. It makes the proportional hazards assumption *within strata.*

Another possibility for a case-subcohort study is for the subcohort to be sampled randomly within strata. This requires that the covariates on which the sampling strata are based are available in the full cohort. One option for analysis of the data arising under this stratified sampling approach is to use a stratified analysis, as in (8.18). If the aim is to perform a stratified analysis then sampling the subcohort within strata may result in some gain in efficiency if certain strata are rare within the population but less so among cases. This might happen if, for example, the strata were determined by sex or age grouping. In Example 8.1 we described a study of lung cancer deaths among aluminium plant workers; in this, the subcohort was in fact sampled within strata defined by decade of birth. Different sampling fractions were used for different decades of birth, so that the distribution of ages in the subcohort approximately matched that in the cases. The analysis was stratified by years of birth divided into five-year periods.

There may be some circumstances in which the subcohort is sampled within strata but the aim is *not* to perform a stratified analysis. In this situation each case is compared with the full case-subcohort risk set in the analysis, not just with those in a particular stratum as in (8.18); that is, this situation is *not* analogous to the use of matching in a nested case-control study. Two particular reasons for following this different approach are the following.

- Stratified subcohort sampling can be used to maximize the informativeness of case-subcohort risk sets, by performing the stratified sampling within strata defined by a confounder of interest that is available in the full cohort. This may be particularly advantageous for estimating interactions between the confounder and a main exposure.
- Another possibility is to implement stratified subcohort sampling using an inexpensive correlate of the main exposure, where the correlate is available in the full cohort. Again, the purpose of this would be to maximize the informativeness of case-subcohort risk sets to gain efficiency.

These examples are analogous to the use of counter-matching in a nested case-control study.

Forming a pseudo-partial likelihood for non-representative stratified sampling raises delicate issues of definition, and we give only brief details about this below.

There appears to be no easy extension of the weighted pseudo-partial likelihood (8.14) to the stratified subcohort sampling context. Suppose that the case at time t_j is outside the subcohort; then what is the probability used in the denominator of (8.14), namely $p(\tilde{S}_j; k, \tilde{\mathcal{H}}_j)$, for an individual k who is *not* the real case? This probability is straightforward to calculate if the real case and individual k are in the same stratum with respect to the covariates used to define the stratified subcohort sampling. But if they are not in the same stratum then the only way for the observed subcohort risk set $\tilde{S}_j$ to have been observed would be if the observed case were swapped with an individual in the relevant subcohort stratum, the latter individual (the 'swapper') then being removed from the comparison set for that case. The swapper would have to be randomly chosen from the relevant stratum. The contributions to (8.14) from all other individuals, including the real case, would be weighted by the inverse probability of their being sampled to the subcohort according to their stratum. This method of weighting results in an unbiased pseudo-score; however, it is cumbersome and difficult to implement.

A simpler alternative weighting strategy for stratified case-subcohort studies uses inverse probability weights, giving a weighted pseudo-partial likelihood of the form

$$\prod_j \frac{\exp\{\beta^T x_{i_j}(t_j)\}\, w_{i_j}(t_j)}{\sum_{k \in S_j \cup C_j} \exp\{\beta^T x_k(t_j)\}\, w_k(t_j)}, \tag{8.19}$$

where, for any individual who is *ever* a case, either within the subcohort or not, the weight is $w_k(t_j) = 1$ and for all other individuals within the subcohort the weight is $w_k(t_j) = n_{s_k}^0(t_j)/m_{s_k}^0(t_j)$, where $n_{s_k}^0(t_j)$ is the total number of 'never-cases' at risk at t_j and in stratum s_k in the full cohort and $m_{s_k}^0(t_j)$ is the number of never-cases at risk at t_j and in stratum s_k in the subcohort. Note that the sum in the denominator is now over $S_j \cup C_j$, where C_j denotes the total set of cases at risk at t_j, that is, it includes cases occurring outside the subcohort at a later time. This weighting scheme results in unbiased score functions at each event time; we refer to appendix section A.7 for further details.

The above illustrates some difficulties involved in the analysis of a case-subcohort study with stratified subcohort sampling. As noted, stratified subcohort sampling on the basis of a surrogate exposure measurement or a fully observed confounder has the same aim as counter-matching in a nested case-control study, that is, to maximize the informativeness of each case-control set. However, this method has some tension with the spirit of case-subcohort studies, a major advantage of the latter being that the same subcohort can be re-used for multiple outcomes; stratified sampling on a surrogate exposure or confounder makes the subcohort specific to the study of a particular investigation.

8.7 Some further approaches

8.7.1 Parametric models

The focus in this chapter has been on the use of the proportional hazards model and semi-parametric pseudo-likelihood analyses. A weighted approach for fitting parametric models in nested case-control studies was outlined in Section 7.8.3. A similar approach is possible in case-subcohort studies. In fact the analysis is simpler in the situation of a case-subcohort study because the probability that an individual is in the case-subcohort sample is 1 for cases and m/N for non-cases in the subcohort, where m is the size of the subcohort and N is the size of the underlying cohort.

8.7.2 Using full-cohort information in case-subcohort studies

It was noted in Section 7.11 that the standard analysis of nested case-control studies is in a sense inefficient because it does not utilize information on covariates or, possibly, on surrogates of main exposures, that may be available for all individuals in the underlying full cohort. The same point applies in the case-subcohort analyses described above. Some information from the full cohort may be used in stratified sampling of the subcohort. However, in many situations appreciable information in the full cohort will be ignored in a case-subcohort analysis.

One way of making better use of full-cohort information is to use 'optimal' weights in the weighted pseudo-partial likelihood (8.19). Two methods for obtaining these optimal weights have been described, resulting in either 'estimated weights' or 'calibrated weights'. Estimated weights are obtained by using full cohort data, including any surrogate exposures, to *predict* who will be sampled to the case-subcohort study. Calibrated weights are obtained by finding weights that are as close as possible to the observed sampling weights, subject to the constraint that the cohort totals of the surrogate exposures are equal to their weighted sums among sampled subjects. We do not give the details here. This method has been found to result in some gains in efficiency relative to the standard case-subcohort analysis.

An alternative way of using full-cohort information is by multiple imputation. In this we impute the large amount of missing data on exposures and covariates for individuals who are in the full cohort but not in the case-subcohort study. This was described in Section 7.11 for nested case-control studies. The issues here are similar and are omitted. An imputation model as in (7.32) may be considered. More complex imputation methods are simpler for a case-subcohort study than for a nested case-control study, because of the presence of a random sample from the full cohort, in the form of the subcohort, which enables sampling from the marginal distribution of exposures.

8.7.3 Estimating absolute risk

Finally, we note that the baseline cumulative hazard can be estimated from a case-subcohort study in a way similar to that described for a nested case-control study, outlined in Section 7.10, by weighting the contribution of individuals in the subcohort by the probability of their being sampled to the subcohort. The baseline cumulative hazard is therefore estimated as

$$\hat{H}_0(t) = \sum_{t_j \leq t} \frac{1}{\sum_{k \in S_j} \exp\{\beta^T(x_k(t) - x_j^*)\}\, n(t_j)/m(t_j)}, \tag{8.20}$$

where $n(t_j)$ denotes the total number at risk in the full cohort at time t_j and $m(t_j)$ is the number at risk in the subcohort at t_j.

8.8 Comparison between nested case-control and case-subcohort studies

The case-subcohort design that makes use of event times, on which we have focused primarily in this chapter, is an alternative to the nested case-control design studied in the previous chapter. Both designs enable the estimation of hazard ratios from a case-control-type sample within a cohort. Both allow for censoring and staggered entry to the underlying cohort and for time-dependent exposures. Both types of study can also be analysed in standard statistical software, despite some complexities of the analyses. Some relative advantages and disadvantages of the two designs are as follows.

- On the one hand, the major advantage of case-subcohort studies within prospective cohorts is that the same subcohort can be re-used in investigating different outcomes, that is, for different case groups. This can bring major savings in cost, depending on the number of exposures of interest that are common across studies. There do exist methods for the re-use of controls in nested case-control studies, but the analysis becomes complex and these methods are not yet well established.
- On the other hand, the matching of controls to specific cases in a nested case-control study enables much finer matching than may be easily achieved in a case-subcohort study using a stratified analysis. In a case-subcohort study some cases with rare matching variables may not have any counterparts in the subcohort, resulting in a loss of information.
- Nested case-control studies are also advantageous if it is important that exposure measurements from a case and its controls are made at the same time, for example if the exposure is a biological measurement that is sensitive to storage time or laboratory batch effects.
- The presence of the subcohort in a case-subcohort study provides the option, were it of interest, to monitor especially closely certain exposures over time in a subset of participants.

Notes

Section 8.2. Kupper *et al.* (1975) described what they referred to as a 'hybrid epidemiologic design', which was the first description of what we have called a case-subcohort study . The idea was also discussed by Miettinen (1976, 1982), who used the term 'case-referent study'. The focus in these papers was on studies *not* using event times. Case-subcohort studies disregarding time have been referred to perhaps more commonly in the literature as 'case-base' studies. The case-subcohort design for use specifically in a survival analysis context using event times was first described by Prentice (1986a), who introduced the term 'case-cohort study'.

Section 8.3. Kupper *et al.* (1975) outlined the estimation of risk ratios in their 'hybrid epidemiological design'. See Greenland (1986) for adjusted risk ratio estimates. See also Miettinen (1982) and Nurminen (1989). A more efficient estimator for the risk ratio is (8.8), given by Sato (1992, 1994); see also Rothman *et al.* (2008, p. 253). Flanders *et al.* (1990) described how risk ratios can be estimated in a case-subcohort study when some individuals are censored before the end of follow-up, using an approach based on life tables that is analogous to methods available for risk ratio estimation in full-cohort studies under similar circumstances.

Section 8.4. The analysis of case-subcohort studies using the pseudo-partial likelihood (8.13) was first outlined by Prentice (1986a), who also derived expressions for the score variance and covariances. Self and Prentice (1988) derived the formal asymptotic properties of the pseudo-maximum-likelihood estimate $\hat{\beta}$ of the log hazard ratio from a case-subcohort analysis.

Section 8.4.2. Self and Prentice (1988) omitted cases outside the cohort from the denominator of their contribution to the likelihood (that is, gave them zero weight) and derived the simpler variance estimator that results under this scheme. An analysis in which subcohort members are weighted by the inverse of the in probability of being sampled to the subcohort (see (8.16)) was suggested by Barlow (1994). Summaries and comparisons of the different weighted pseudo partial likelihood analyses are given in Barlow *et al.* (1999) and Onland-Moret *et al.* (2007). The definition and derivation of influence functions for proportional hazards regression were given by Reid and Crépeau (1985), with later discussion by Barlow and Prentice (1988). A robust estimator for case-subcohort studies was developed by Barlow (1994). See also Lin and Ying (1993), who discussed variance estimation in the context of considering the case-subcohort design as a full cohort study with missing data. See Therneau and Li (1998) for a discussion of the alternative estimators and their implementation in statistical software packages. Wacholder *et al.* (1989) outlined a way of bootstrapping in case-subcohort studies to obtain variance estimates.

Section 8.5. The asymptotic relative efficiency of a case-subcohort study relative to a full-cohort study was derived by Self and Prentice (1988). The effects of the size of the subcohort were studied by Onland-Moret *et al.* (2007) using simulation studies. Barlow *et al.* (1999) commented on the possibility of augmenting the subcohort if its size becomes too small at later event times.

Section 8.6. Borgan *et al.* (2000) extended the case-subcohort design to the stratified subcohort sampling situation, proposing three different possibilities for the weighted pseudo-partial likelihood. These included the 'swapper' approach described in this section, and the version given in (8.19); the latter is sometimes referred to as the Borgan II estimator. The third weighting approach omits all cases outside the subcohort from the denominator terms and weights all other individuals, including cases, by $w_k(t_j) = n_{s_k}(t_j)/m_{s_k}(t_j)$, where $n_{s_k}(t_j)$ is the total number of individuals at risk at t_j and in stratum s_k in the *full cohort* and $m_{s_k}(t_j)$ is the number of individuals at risk at t_j and in stratum s_k in the *subcohort*. These three different approaches were found to perform similarly by Borgan *et al.* (2000). They derived asymptotic results and a variance estimator for the stratified situation. See also Samuelsen *et al.* (2007).

Section 8.7.1. The use of a weighted pseudo-likelihood to fit fully parametric models in case-subcohort studies was outlined by Kalbfleisch and Lawless (1988), who also gave a more general discussion of the use of weighted pseudo-likelihoods in studies of disease incidence with a focus on studies with a retrospective element, including multi-state situations.

Section 8.7.2. Optimally weighted analyses were outlined by Breslow *et al.* (2009b), in the context of a stratified case-subcohort study though the method applies more generally. This was related to earlier work by Kulich and Lin (2004). See also Breslow *et al.* (2009a). The use of multiple imputation to include full-cohort data in case-subcohort studies was described by Marti and Chavance (2011) and Keogh and White (2013). There are also a number of methods arising from the literature on how to deal with missing data in proportional hazards regression that could be applied to the case-subcohort situation, as a special case, as a way of using full-cohort information. See, for example, Prentice (1982), Zhou and Pepe (1995), Nan (2004) and Chen (2002). These methods do not appear to have been often used in applications.

Section 8.8. Langholz and Thomas (1990) discussed the practicalities of the nested case-control and case-subcohort designs and compared their efficiency under different circumstances.

9

Misclassification and measurement error

- The misclassification of exposures and outcomes and errors in continuous exposures result in biased estimates of associations between exposure and outcome. A particular consideration that arises in case-control studies is differential error or misclassification that depends on the outcome.
- Relatively simple methods can be used to correct for misclassification in binary exposures, provided that there is information available on the sensitivity and specificity of the measured exposure, for example from a validation study. These methods extend to allow differential misclassification and additionally to allow for misclassification in binary outcomes.
- Error in continuous exposures arises in many areas of application and can take different forms. The form of the error influences its effect on the estimated association between exposure and outcome.
- A commonly used method for correcting error in continuous exposures is regression calibration, which relies on an assumption of non-differential error. Correction methods that allow differential error include multiple imputation and moment reconstruction.

9.1 Preliminaries

In this chapter we discuss the effects of misclassification and measurement error and methods for making corrections for these effects. The focus is naturally on case-control studies, but much of the discussion and methods apply more generally. After some preliminary remarks, the chapter is divided broadly into three sections:

- misclassification of a binary or categorical exposure;
- misclassification of case-control status;
- error in the measurement of a continuous exposure.

Measurement error may be a major issue in many areas of investigation and types of study. In general, error may occur in the ascertainment of both exposures and outcomes. However, error in exposure measurements is usually of more concern.

Error in exposure measurements can arise for a number of reasons, including

- imprecision of a diagnostic test used to measure an exposure;
- laboratory error in measurements of exposures that are biological;
- the self-reporting of exposures, for example in the number of cigarettes smoked per day;
- when the true exposure of interest is a long-term average level, for example that of blood pressure, but it has been possible to obtain only one or a small number of measurements on each individual.

The retrospective character of the collection of exposure information in some case-control studies means that the exposure information obtained may be prone not only to potentially greater error than had the information been collected prospectively but also to specific types of error, notably the differential misclassification of exposures or differential error in continuous exposures by case or control status. This may arise owing to differential recall among cases and controls, where cases have perhaps dwelt on the potential causes of their diagnosis. Another situation in which differential error could arise is where exposure information for cases is obtained from a proxy, for example a relative, if the case individual is too ill to respond or is dead. For a biological exposure it is plausible that case status or perhaps the use of certain medications may affect the quality of a laboratory measurement. Differential error is therefore a special concern in some case-control studies and efforts to correct for it are often required. Some case-control studies are nested within larger prospective studies within which exposure information has been collected prospectively, before the outcome occurred. In these differential error is of no more concern than it would be in the corresponding full-cohort study.

There is a substantial literature on methods for correcting for measurement error.

Example 9.1: Assessment of differential error   Zota *et al.* (2010) investigated the association between breast cancer risk and certain chemicals found in household cleaners, air fresheners and pesticides, using a population-based case-control study. Exposure information was gathered retrospectively by telephone interview. All participants were asked about their use of products containing pesticides in the home, for example insect repellent, and a subset were asked about household cleaning products. Information obtained included whether they had ever used a given product, the frequency of use and the time period over which the use occurred.

The investigators in this study were concerned about differential recall bias among cases and controls. To enable some study of such potential bias, as part of the telephone interview participants were asked to what degree they believed certain factors, including chemicals and air pollutants, were associated with breast cancer risk.

The authors looked for evidence of differences in beliefs between cases and controls and for interactions in exposure–breast-cancer associations and beliefs. They performed analyses stratified by belief or otherwise in an association. There was a small but statistically significant difference between the proportion of cases (60%) and controls (57%) who expressed a belief that chemical and pollutants were associated with breast cancer. This study found a strong positive association between cleaning product use and breast cancer, but no association with pesticide use. However, the odds ratios were consistently higher in the subset who believed there was an association, and this was observed for both cleaning products and pesticides.

9.2 Misclassification in an exposure variable

9.2.1 Describing misclassification

Binary or categorical exposure variables may be subject to misclassification. Here we focus on a case-control study with a single binary exposure subject to misclassification. The *true exposure* is denoted X_t and the *measured exposure* is denoted X_m. Misclassification occurs when $X_t \neq X_m$. The binary outcome Y that is the basis of the case-control sampling is for now assumed to be observed without error. The assumption, which may not always be realistic, is that there are stable probabilities

$$\Pr(X_m = 1 | X_t = 1, Y = y) = 1 - \epsilon_X, \tag{9.1}$$

$$\Pr(X_m = 1 | X_t = 0, Y = y) = \eta_X. \tag{9.2}$$

In this simplest formulation misclassification is assumed to occur independently of disease status Y and we have shown this explicitly in the notation. In the context of diagnostic testing, $1 - \epsilon_X$ and $1 - \eta_X$ are, respectively, the *sensitivity* and *specificity* of the procedure. The implicit assumption is that ϵ_X and η_X are characteristics of the measuring process and do not depend on the particular population under study.

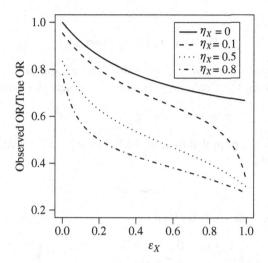

Figure 9.1 Effects of misclassification on odds ratio estimates; the true odds ratio was 2.

Misclassification of the form described above results in an attenuation of the estimated odds ratio, that found using X_m, compared with the true odds ratio, that found using X_t. We denote the true odds ratio, obtained using X_t, by ϕ. Specific situations can be investigated numerically. For example, suppose that in the controls $Pr(X_t = 1|Y = 0) = 1/2$; then the estimated odds ratio using X_m can be written

$$\frac{\{(1 - \epsilon_X)\phi + \eta_X\}(1 - \eta_X + \epsilon_X)}{(\epsilon_X\phi + 1 - \eta_X)(1 - \epsilon_X + \eta_X)}. \tag{9.3}$$

If the error probabilities are small then this ratio becomes

$$\phi\{1 - (\phi - 1)(\epsilon_X + \eta_X/\phi)\}. \tag{9.4}$$

Figure 9.1 shows in more detail the effects of misclassification on odds ratio estimates.

9.2.2 Correction for misclassification

We now outline some methods for correcting for the effects of misclassification. We start with a population model described by probabilities

$$Pr(X_t = x, Y = y) = \pi_{xy,t} \tag{9.5}$$

and

$$\Pr(X_m = x, Y = y) = \pi_{xy,m} \qquad (9.6)$$

where the suffices t, m distinguish the true and measured values. We may consider these as 4×1 column vectors π_m, π_t, where, for example,

$$\pi_t = (\pi_{00,t} \quad \pi_{01,t} \quad \pi_{10,t} \quad \pi_{11,t})^T. \qquad (9.7)$$

Then, by the laws of probability, in the light of the independence assumptions specified above,

$$\pi_m = A\pi_t \qquad (9.8)$$

where the 4×4 matrix A is given by

$$\begin{pmatrix} 1 - \eta_X & 0 & \epsilon_X & 0 \\ 0 & 1 - \eta_X & 0 & \epsilon_X \\ \eta_X & 0 & 1 - \epsilon_X & 0 \\ 0 & \eta_X & 0 & 1 - \epsilon_X \end{pmatrix}. \qquad (9.9)$$

The true odds ratio is $\phi = (\pi_{11,t} \, \pi_{00,t})/(\pi_{10,t} \, \pi_{01,t})$. By using the result $\pi_t = A^{-1}\pi_m$ it can be shown that, given estimates of ϵ_X, η_X, the parameter ϕ can be estimated in the case-control data with mismeasured exposure using

$$\frac{\{\Pr(X_m = 1|Y = 1) - \eta_X\}\{\Pr(X_m = 0|Y = 0) - \epsilon_X\}}{\{\Pr(X_m = 1|Y = 0) - \eta_X\}\{\Pr(X_m = 0|Y = 1) - \epsilon_X\}}. \qquad (9.10)$$

One possibility is that suitable estimates of (ϵ_X, η_X) are available from an external study. Alternatively, a validation study may be conducted within the case-control sample. For some exposures it will never be possible to observe X_t. It is assumed that the error probabilities are stable properties of the measurement process, unaffected by the frequencies of various outcomes encountered and by the method of sampling used and thus, for example, reasonably estimated from replicate determinations of X_m. In general, at least three replicates are required to estimate the misclassification probabilities.

If the error probabilities ϵ_X, η_X are known or have been estimated with negligible error then an estimate of the variance of the corrected log odds ratio estimate can be obtained in the usual way; see Section 2.3.2. If, however, the estimates of ϵ_X, η_X are subject to non-negligible error then this additional uncertainty should be accounted for in the variance of the corrected log odds ratio estimate. Where (ϵ_X, η_X) are estimated in a validation study, the variance of the final estimates can be obtained directly, for example by local linearization, the so-called delta method. An alternative is to estimate the variance of the corrected odds ratio by bootstrapping. To do this, samples within the case and control groups would be taken separately within the validation subsample and within the remainder of the case-control data.

9.2.3 Differential misclassification

Above we focused on non-differential misclassification, that is, on misclassification probabilities that do not differ according to Y. If there is differential misclassification of the exposure by case or control status then we define

$$Pr(X_m = 1 | X_t = 1, Y = y) = 1 - \epsilon_{Xy},$$
$$Pr(X_m = 1 | X_t = 0, Y = y) = \eta_{Xy}.$$

The above procedures can be extended in an obvious way to this situation and we do not give the details here. In this situation any validation study used to estimate misclassification probabilities would have to include both cases and controls.

9.2.4 A full-likelihood approach

The above methods for correcting for exposure misclassification do not accommodate adjustment for covariates W, say. For this, a likelihood analysis may be considered.

We suppose now that a validation sample is taken within the case-control study and that within this validation sample both X_t and X_m are observed; in the remainder of the study only X_m is observed. The outcome Y and adjustment variables W are observed for all individuals and, it is assumed, without error.

We assume that the W are categorical here for ease of notation, but extensions to continuous covariates follow directly. The contribution to the likelihood from an individual in the validation sample is

$$Pr^{\mathcal{D}}(X_t = x_t, X_m = x_m, W = w | Y = y) \tag{9.11}$$

and the contribution from an individual *not* in the validation sample is

$$\sum_{x_t} Pr^{\mathcal{D}}(X_t = x_t, X_m = x_m, W = w | Y = y). \tag{9.12}$$

The probabilities in (9.11) and (9.12) can be written in the form

$$\Pr^{\mathcal{D}}(X_m = x_m | X_t = x_t, Y = y, W = w)$$
$$\Pr^{\mathcal{D}}(Y = y | X_t = x_t, W = w)\Pr^{\mathcal{D}}(X_t = x_t, W = w)/\Pr^{\mathcal{D}}(Y = y).$$

$$(9.13)$$

The second factor is of primary interest and its familiar logistic formulation is given by

$$\Pr^{\mathcal{D}}(Y = y | X_t = x_t, W = w) = L_y(\alpha^* + \beta_{x_t} + \gamma_w^T). \qquad (9.14)$$

The first factor in (9.13) is the conditional misclassification probability and a logistic or multi-logistic model may be chosen. If the misclassification is assumed to be independent of W then simplifications can be made. A suitable model may be chosen for $\Pr^{\mathcal{D}}(X_t = x_t, W = w)$, and $\Pr^{\mathcal{D}}(Y = y)$ may be assumed to be a constant. The full likelihood is the product of the probabilities in (9.11) over individuals in the validation sample multiplied by he product of the sum of probabilities in (9.12) over individuals in the remainder of the case-control sample. Parameter estimates can be obtained by maximum likelihood.

9.3 Misclassification of cases and controls

The methods described in Section 9.2.1 can be extended to incorporate the misclassification of cases and controls. We denote the true outcome by Y_t and the measured outcome by Y_m, and define, omitting the superscript $\mathcal{D}$,

$$\Pr(Y_m = 1 | Y_t = 1) = 1 - \epsilon_Y, \quad \Pr(Y_m = 1 | Y_t = 0) = \eta_Y.$$

In this case the misclassification matrix A in (9.8) becomes

$$A_X \otimes A_Y, \qquad (9.15)$$

where $\otimes$ denotes a Kronecker product of matrices and, for example,

$$A_X = \begin{pmatrix} 1 - \eta_X & \epsilon_X \\ \eta_X & 1 - \epsilon_X \end{pmatrix}, \qquad (9.16)$$

where $1 - \epsilon_X$ and $1 - \eta_X$ are the sensitivity and specificity of the observed exposure measurements as defined in (9.1) and (9.2). Note that here we assume non-differential misclassification in the exposure. It follows from (9.8) that

$$\pi_t = A^{-1}\pi_m, \qquad (9.17)$$

where

$$A^{-1} = A_X^{-1} \otimes A_Y^{-1}. \qquad (9.18)$$

As before, given independent estimates of the error rates, (9.18) can be used to correct for misclassification estimates of odds ratios obtained from measured values in order to give an estimate of the true odds ratio.

To show the general implications of misclassification and, in particular, the implications for case-control studies, we suppose that the error probabilities are small and ignore expressions quadratic in (ϵ, η). For example, we replace $(1 - \eta_X)(1 - \eta_Y)$ by $1 - \eta_X - \eta_Y$. Then

$$A = I + B + O\{(\epsilon, \eta)^2\}, \quad A^{-1} = I - B + O\{(\epsilon, \eta)^2\},$$

where I is the identity matrix and

$$B = \begin{pmatrix} -\eta_X - \eta_Y & \epsilon_Y & \epsilon_X & 0 \\ \eta_Y & -\eta_X - \epsilon_Y & 0 & \epsilon_X \\ \eta_X & 0 & -\epsilon_X - \eta_Y & \epsilon_Y \\ 0 & \eta_X & \eta_Y & -\epsilon_X - \epsilon_Y \end{pmatrix}. \qquad (9.19)$$

It follows, after some calculation, that the probability limit of the ratio of the odds ratio from the measured values and that from the true values is, to the order considered,

$$1 + \epsilon_X(\pi_{10}/\pi_{00} - \pi_{11}/\pi_{01}) + \epsilon_Y(\pi_{01}/\pi_{00} - \pi_{11}/\pi_{10})$$
$$+ \eta_X(\pi_{01}/\pi_{11} - \pi_{00}/\pi_{10}) + \eta_Y(\pi_{10}/\pi_{11} - \pi_{00}/\pi_{01}). \qquad (9.20)$$

It remains to interpret the probabilities π_{xy} entering this formula, and here the method of sampling is important; for case-control sampling the probabilities have to be taken conditionally on an individual's entering the data. To the order considered the probabilities may refer to the true or to the measured values.

The coefficients of (ϵ_X, η_X) have a direct interpretation for case-control sampling and therefore can be directly estimated. For example, the coefficient of ϵ_X in (9.20) is

$$\frac{\Pr(X = 1 | Y = 0)}{\Pr(X = 0 | Y = 0)} - \frac{\Pr(X = 1 | Y = 1)}{\Pr(X = 0 | Y = 1)}; \qquad (9.21)$$

for the coefficient of η_X the roles of 0 and 1 are interchanged.

For the coefficients of (ϵ_Y, η_Y), the roles of X and Y are interchanged. For interpretation in the case-control context, however, we use, wherever possible, conditional probabilities of X given Y, so that the coefficient of ϵ_Y in (9.20) is

$$\frac{\Pr(Y=1)}{\Pr(Y=0)}\left\{\frac{\Pr(X=0|Y=1)}{\Pr(X=0|Y=0)}-\frac{\Pr(X=1|Y=1)}{\Pr(X=1|Y=0)}\right\}. \qquad (9.22)$$

The coefficient of η_Y is obtained by interchanging the 0s and 1s.

This requires information about the ratio of probabilities $\Pr(Y=1)/\Pr(Y=0)$ in addition to the conditional probabilities, which can be estimated as before. Some care is needed in specifying $\Pr(Y=1)/\Pr(Y=0)$. The normal procedure would be that, as each case is found, to sample one or more, say c, controls. This suggests taking $\Pr(Y=1)/\Pr(Y=0)$ to be $1/c$. Note, however, that if in fact cases are found by examining a large number of possibilities, then $\Pr(Y=1)$ is quite small for each possibility examined and this implies that the coefficient of η_Y, which includes a term in $1/\Pr(Y=1)$, may be large. In this situation it is important that η_Y is very small. Indeed, if cases are rare and controls common, even a small probability that a control is classified as a case could lead to a substantial proportion of false cases, severely attenuating any differences that are present. This stresses the importance of a clear protocol for the definition of a case.

9.4 Misclassification in matched case-control studies

We have so far considered unmatched case-control studies, and we now move on to consider briefly some corresponding methods for matched studies. For simplicity the focus is on matched-pair studies. Let $X_{t;1k}$ and $X_{t;0k}$ denote the true exposures of the case and the control respectively in the kth matched pair, and let $X_{m;1k}$, $X_{m;0k}$ be the corresponding measured exposures. It is assumed that misclassification probabilities do not differ by matched set but may differ by case-control status. In our earlier notation, and denoting the matched set by k, the misclassification probabilities are given as

$$\Pr(X_{m;yk}=1|X_{t;yk}=1)=1-\epsilon_{Xy}, \quad y=0,1,$$
$$\Pr(X_{m;yk}=1|X_{t;yk}=0)=\eta_{Xy}, \quad y=0,1.$$

The number of matched pairs in which $(X_{t;1k}=i, X_{t;0k}=j)$ is denoted by $n_{t;ij}$, and the corresponding observed number of matched pairs with $(X_{m;1k}=i, X_{m;0k}=j)$ is denoted by $n_{m;ij}$. The expected number of matched pairs of the four types

$$n_{m;11}, \quad n_{m;10}, \quad n_{m;01}, \quad n_{m;00}$$

can be expressed in terms of the misclassification probabilities and the $n_{t;ij}$ using

$$n_m = An_t, \tag{9.23}$$

where $n_m = (n_{m;00} \ n_{m;01} \ n_{m;10} \ n_{m;11})^T$, and similarly for n_t, and the matrix A is

$$A = A_0 \otimes A_1,$$

where

$$A_y = \begin{pmatrix} 1 - \eta_{X1} & \epsilon_{X1} \\ \eta_{X1} & 1 - \epsilon_{X1} \end{pmatrix}, \quad y = 0, 1.$$

It can be shown using the above results that the corrected odds ratio is

$$\frac{n_{m;10}\eta'_{X1}\epsilon'_{X0} + n_{m;01}\epsilon_{X0}\eta_{X1} - n_{m;11}\epsilon_{X0}\eta'_{X1} - n_{m;00}\epsilon'_{X0}\eta_{X1}}{n_{m;10}\epsilon_{X1}\eta_{X0} + n_{m;01}\epsilon'_{X1}\eta'_{X0} - n_{m;11}\epsilon_{X1}\eta'_{X0} - n_{m;00}\epsilon'_{X1}\eta_{X0}}, \tag{9.24}$$

where $\epsilon'_{Xy} = 1 - \epsilon_{Xy}$ and $\eta'_{Xy} = 1 - \eta_{Xy}$.

A likelihood-based approach provides a more general way of correcting for exposure misclassification in a matched case-control study. Here we define $n_{m;ij}$ to be the number of observed matched sets with i exposed cases and j exposed controls. The $\{n_{m;ij}\}$ come from a multinomial distribution with probabilities $\{\omega_{m;ij}\}$, say, and the full likelihood is

$$\text{lik} = \prod_{ij} \omega_{m;ij}^{n_{m;ij}}. \tag{9.25}$$

In the simplest case of $1 : 1$ matching, the probabilities $\omega_{m;ij}$ can be formulated in terms of the sensitivity and specificity (which may differ for cases and controls) and the true odds ratio parameter. The above likelihood could be extended to include parts referring to a validation subset consisting of case-control sets and the remainder of the sets, as in the unmatched situation in Section 9.2.4.

9.5 Error in continuous exposures

9.5.1 Preliminaries

We now proceed to consider errors in the measurements of continuous exposures. In this situation the definition of the true exposure X_t may correspond, for example, to a gold-standard method of measurement or to a time average of a stable but erratically varying quantity. Errors in continuous exposures can take many forms. It is often helpful to regard measurement errors as having both a systematic and a random component. We focus primarily here on random errors, though systematic errors will be touched upon.

There are a number of important distinctions to be made in considering the random component of variability. For example, the error may be multiplicative or additive. Another important distinction is that between the so-called classical error model and the Berkson model.

- Under the *classical error model* the measured exposure X_m can be expressed as the true exposure X_t plus a random variable, ϵ say, which is independent of X_t. In an extension, ϵ may have a conditional mean equal to zero for all values of x_t but a distribution depending on x_t, for example having a large variance at the extremes of the range of measurement.
- In the *Berkson error model*, by contrast, the error term is independent of X_m.

There are intermediate cases. The classical model is the most commonly appropriate, although both models or a combination may arise in a case-control context. Below, we discuss further the classical and Berkson error models and their implications.

9.5.2 Classical error model

The classical error model is

$$X_m = X_t + \epsilon, \tag{9.26}$$

where ϵ is a random variable of mean zero and variance σ_ϵ^2 distributed independently of X_t and independently for different study individuals.

When X_m is used in place of X_t in an investigation, this model has the following consequences.

- A simple linear regression relation with one explanatory variable is attenuated, that is, flattened; see Figure 9.2. Attenuation will also typically be the effect of classical error in non-linear regression models, including the logistic model.
- For multiple regression the situation is more complicated. Explanatory variables that have relatively large measurement error may have their dependence in a sense shifted to correlated variables that are well measured

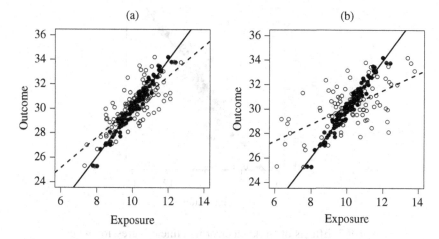

Figure 9.2 Effects of classical measurement error in a linear regression: (a) the variance of the error is 0.5 times the variance of the true exposure; (b) the variance of the error is 1.5 times the variance of the true exposure. Solid circles and lines, the true values. Open circles and broken lines, values with measurement error. Regression lines are shown.

but are in fact unimportant for explanatory purposes. That is, while there is a general attenuation some apparent dependences may be increased.

- There may be some distortion of the shape of the relation in regressions involving non-linear terms in X_m.

To assess the effect of classical measurement error, and to correct for it, estimates are required of the measurement error variance σ_ϵ^2. This can be estimated in a validation study within which the true exposure X_t is observed. If it is not possible to observe the true exposure then the error variance may be estimated using one or more repeated exposure measurements X_m on at least a subset of individuals. Perhaps less commonly, an estimate of the error variance may be available from an external study. A summary of correction methods follows in a later section.

9.5.3 Berkson error

The assumption formulated in the classical error model in (9.26), that the error is statistically independent of the true value, will often be reasonable, for example in measuring log blood pressure. It is the basis of most discussion in the remainder of this chapter. An extreme contrast is with

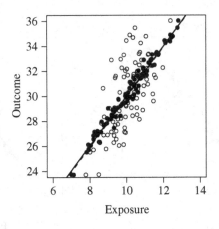

Figure 9.3 Effects of Berkson error in a linear regression when the variance of the error term ϵ^B is 1 and the variance of X_m is 1. Solid circles and lines, the true values; open circles and broken lines, values having measurement error.

situations with so-called Berkson error, in which the error is independent of the measured value. This arose first in an experimental situation but we discuss it here in an observational setting.

Note first that for the classical error model the measured values are more variable than the true values. In the Berkson model the reverse is true, and its simplest form, corresponding to (9.26), is

$$X_t = X_m + \epsilon^B, \qquad (9.27)$$

where ϵ^B has zero mean and is independent of X_m. In the simplest case, that of linear regression, the effect of Berkson error is to increase the variance about the regression line but to leave the regression line unchanged; see Figure 9.3. In a logistic regression, however, Berkson error can result in some attenuation of the association.

Berkson error arises in contexts such as the following. Suppose that the study individuals are distributed in space and or time and that it is not possible to measure the exposure for each individual but space and or time averages are available. The true value for each individual may then be assumed to be the local spatial average plus an individual deviation corresponding to ϵ^B, and this is essentially the Berkson formulation.

Example 9.5: Berkson error Reeves *et al.* (1998) described methods for measurement error correction motivated by case-control studies of residential exposure to radon and the risk of lung cancer. The focus was on the average exposure to radon in the home over many years. Two possible ways in which the exposure may be measured are:

(1) by taking for each individual a radon measurement in homes where they have lived over the relevant time period;
(2) if the previous home cannot be accessed, then the radon measurement may be taken instead from an average of nearby locations.

Measurements of the first type may be reasonably assumed to follow the classical error model. For those of the second type, however, a Berkson error model is appropriate. Because in this second case the measurement obtained from the nearby location is also likely to carry an error, a modification of the Berkson model may be used. Reeves *et al.* (1998) considered the following error representations:

$$X_m = \tilde{X} + \epsilon_m, \quad X_t = \tilde{X} + \epsilon_t. \tag{9.28}$$

This representation captures a mixture of both classical and Berkson error.

9.5.4 Extensions to the classical error model

We now consider some more complicated possibilities for the form of the error, to be represented therefore by different models. The focus is on extensions to the classical form of error and not to Berkson error.

One possibility is that error may be of different importance in different parts of the range of exposure values; for example, the error may be multiplicative rather than additive. The error model in (9.26) is extended to this situation by allowing ϵ to have conditional mean zero for all values of X_t but a distribution depending on X_t.

Under the classical error model and its extension to multiplicative error, the measured exposure X_m is described as an unbiased measure of the true exposure X_t because the average over a large number of repeated measurements of X_m would provide an estimate of X_t. In many situations this will be an appropriate model, for example for biological measurements that are subject to random laboratory error or to fluctuations over time (for example blood pressure or residential radon exposure). There are some situations, however, where this may be an unrealistic assumption. Although it is desirable to eliminate systematic errors at the measurement stage, this is not always possible. One formulation for systematic error is

$$X_m = \theta_0 + \theta_1 X_t + \epsilon, \qquad (9.29)$$

where the errors ϵ have mean 0 and constant variance. The classical error model is a special case of model (9.29) with $\theta_0 = 0, \theta_1 = 1$. Correction for systematic errors is possible in some situations, including

- when the form of the systematic error is known from an external study, although this may be an uncommon situation;
- when the true exposure X_t is observed in a validation sample within the main study sample;
- when a 'superior' measure, X_s say, is available in a validation sample that is still error prone but is an unbiased measure of the true exposure, for example, because it follows the classical error model.

Finally, error may be differential in the sense that for cases it has a distribution different from that for controls. This extension is of particular relevance in case-control studies. There are a number of forms that differential error can take. A very simple possibility is that the mean of the errors ϵ in the classical error model differs by a constant in cases and in controls. Another simple model for differential error extends the classical error model by allowing error variances to differ in the case and control groups. In order to understand and correct for the effects of differential error, information must be available on the error in both the case and the control groups. Depending on the situation and on the precise form of the error, this information may come from an external source, from a validation study or from repeated exposure measurements on some individuals.

Both systematic and differential errors in X_m may have important consequences for interpretation, and estimates of associations with the outcome may be biased away from or towards the null association. Particular care is needed if attention is focused on the extremes of an exposure range.

Example 9.4: Error in dietary intake measures In a matched case-control study of the association between fibre intake and colorectal cancer within the EPIC-Norfolk cohort, introduced in Example 3.1, dietary intake was measured using information obtained from seven-day diet diaries, which were collected at recruitment to the underlying cohort although not processed until later. Measures of dietary intake are known to be subject to potentially substantial errors. The exposure of interest is 'long-term average' intake, while seven-day diet diaries record short-term intake. Error

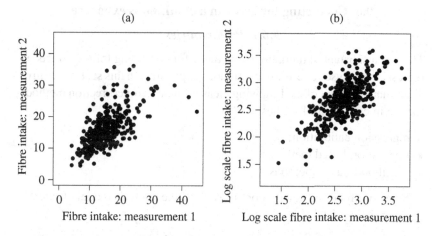

Figure 9.4 Plots of replicate measurements (1 and 2) of fibre intake in the EPIC-Norfolk study: (a) on the original scale; (b) on a log scale.

therefore arises due to the natural fluctuations in dietary intake over time. Other sources of error include the misreporting of food items and of portion sizes and errors at the data-processing stage where food intake is translated into nutrient intake.

The prospective collection of diet diaries in this study is advantageous because it substantially reduces fears of the differential reporting of dietary intake by cases and controls. Individuals who received a colorectal cancer diagnosis within 12 months of completing their diet diary were excluded. Repeated seven-day diary measurements were available for a subset of participants, and these allow some assessment of measurement error.

Figure 9.4 shows plots of the two fibre measurements on the original scale and on a log scale. Using the original-scale measurements, there appears to be an increase in error variance as the measurement values increase. On the log scale, the error variance appears approximately constant. It may be appropriate here to assume the classical error model, as in (9.26), using the log scale measurements.

There is a concern about systematic error in dietary measurements, and a model such as that in (9.29) may be more appropriate. This suggestion arises from comparison not with the true exposure, which cannot really be observed, but by comparison with biomarkers available for some nutrients which are believed to provide unbiased measures of exposure.

9.6 Correcting for error in a continuous exposure

9.6.1 Preliminaries

There is a substantial literature on methods for correcting for the effects on regression analyses of error in continuous exposures. In this section we give a summary account of the following measurement-error correction methods for use in case-control studies:

- regression calibration;
- imputation-based methods;
- likelihood-based methods.

Since our focus is on case-control studies, the analysis model of interest is the logistic model

$$\Pr{}^{D}(Y = 1 | X_t = x_t) = L_1(\alpha^* + \beta_t x_t). \tag{9.30}$$

If the true exposure X_t is replaced by the measured exposure X_m in the above then a biased estimate of β_t is obtained. This biased estimate is sometimes referred to as the *naive estimate*.

Our focus will be mainly on the classical measurement error model in (9.26), though extensions to other types of error, in particular differential error, which is of particular relevance for case-control studies, are also noted.

9.6.2 Regression calibration

Regression calibration is a widely used method for correcting for error in continuous exposure measurements. In this method the true exposure X_t is replaced in the exposure-outcome model of interest by its expectation conditional on the observed exposure, that is, by $E(X_t | X_m = x_m)$. The popularity of this method arises from its relative simplicity and its applicability in a number of different situations. The use of regression calibration does, however, rest on approximations and assumptions that are particularly critical for non-linear models and may give misleading results, especially if exposures at the end of the range of interest are involved. The joint distribution of true and measured values must be reasonably well specified for the conditional expectation to be found. Here we discuss the use of regression calibration in a relatively simple case-control context, assuming the logistic model (9.30). We note some extensions in a later section.

The left-hand side of the logistic model (9.30) is $E(Y | X_t - x_t)$. The expectation of Y, conditional on what is observed in the case-control sample, is

$$E(Y|X_m) = E\{E(Y|X_m, X_t)|X_m\}. \tag{9.31}$$

Under an assumption that the error in X_m is independent of the outcome Y, that is, the error is *non-differential*, the above can be written as

$$E(Y|X_m) = E\{E(Y|X_t)|X_m\}$$
$$= E\{L_1(\alpha^* + \beta_t X_t)|X_m\}. \tag{9.32}$$

If the regression of Y on X_t were linear, so that $E(Y|X_t) = \omega_0 + \omega_1 X_t$, say, then the above equation would give $E(Y|X_m) = \omega_0 + \omega_1 E(X_t|X_m)$; that is, the parameters in the regression model of interest can be estimated by replacing X_t by its expectation conditional on what is observed. For a logistic regression the calculation is more complicated, however.

The simplest classical error model is specified by defining X_m as being a simple function, with true value, X_t, plus error. If X_t is regarded as a random variable then we can consider the linear least squares regression equation of X_t on X_m, namely

$$X_t = \lambda_0 + \lambda_{t|m} X_m + \zeta. \tag{9.33}$$

Here the error ζ is uncorrelated with X_m. In the special case where the joint distribution of (X_t, X_m) is a bivariate normal distribution, (9.33) is a full regression relation and ζ is independent of X_m and in fact is itself normally distributed.

The expectation in (9.32) cannot be evaluated in closed form but for most purposes a very good approximation can be obtained by first replacing $L_1(x)$ by a suitable multiple, $\Phi(kx)$, of the standard normal integral, where k is a suitably chosen constant. A similar approximation was noted in Section 4.4. Usually $k = (1.7)^{-1}$ is adequate. The integral can then be evaluated and converted back to logistic form to give

$$E(Y|X_m) \approx L_1\left\{\frac{\alpha^* + \beta_t \mu_{X_t|X_m}}{(1 + k^2 \sigma^2_{X_t|X_m} \beta_t^2)^{1/2}}\right\}, \tag{9.34}$$

where $\mu_{X_t|X_m}$ and $\sigma^2_{X_t|X_m}$ respectively denote the mean and variance of X_t given X_m. Under the above assumptions, $\mu_{X_t|X_m}$ and $\sigma^2_{X_t|X_m}$ can be estimated from the linear regression (9.33).

Hence we have shown that when the distribution of X_t conditional on X_m, $X_t|X_m = x_m$, is normally distributed, $\Pr(Y = 1|X_m = x_m)$ can be represented approximately in logistic regression form:

$$\Pr(Y = 1 | X_m = x_m) = L_1(\alpha_m^* + \beta_m x_m), \tag{9.35}$$

where β_m is approximately

$$\beta_m = \frac{\beta_t \lambda_{t|m}}{(1 + k^2 \sigma_{X_t|X_m}^2 \beta_t^2)^{1/2}}. \tag{9.36}$$

When $\sigma_{X_t|X_m} \beta_t$ is small the denominator will be close to unity. The above method then reduces to the replacement of X_t by $\mu_{X_t|X_m}$ in the logistic regression or, equivalently, to estimating β_t using $\beta_m / \lambda_{t|m}$. This simpler method is widely used. It is exact in the case of a linear regression. It has been found that the same approximation can also be used in the setting of a proportional hazards model, making this approach also suitable for use in nested case-control and case-subcohort studies using partial-likelihood analyses.

At this point it is useful to discuss how to obtain estimates of $\mu_{X_t|X_m}$ and $\sigma_{X_t|X_m}^2$ in a case-control study, that is, estimates to fit the model (9.33). One perhaps unusual possibility is that there is external information on the linear regression relation for $X_t | X_m$. Otherwise it is necessary to obtain additional information from within the case-control sample. For rare outcomes the distribution of $X_t | X_m$ will be approximately the same among the controls as it is in the population. Hence a validation sample could be taken within the control group in the case-control study, where X_t is obtained alongside X_m. There are some situations in which it is not possible to observe the 'true' exposure. In this case, under the classical measurement error model for X_m in (9.26) the distribution of $X_t | X_m$ can be estimated if repeated measurements of X_m are available.

The uncertainty in estimation of the parameters in (9.33) should in principle be carried over into the estimate of the variance of corrected log odds ratio $\hat{\beta}_t$. This may be done using local linearization or the delta method or by bootstrapping. In practice the uncertainty in estimation of measurement error parameters is often ignored and may indeed be small in many situations.

When the logistic regression model includes adjustment for additional variables W, then the expectation $E(X_t | X_m = x_m)$ must be replaced by $E(X_t | X_m = x_m, W = w)$ in the above formulations. In special cases we may write $X_t = \lambda_0 + \lambda_{t|m} X_m + \lambda_W W + \zeta$. In a matched case-control study it is appropriate to include the matching variables in W where possible.

The use of regression calibration under a logistic, or other non-linear, exposure-outcome model in a retrospective case-control setting relies on a number of assumptions. If the outcome is not particularly rare or if X_t has an appreciable exposure effect on the outcome, estimating the population distribution of X_t given X_m using a validation sample within the control group will result in biased estimates.

Regression calibration depends crucially on an assumption that the error in X_m is non-differential, that is, it does not differ according to case or control status. As noted earlier, error that is differential by case-control status is often a concern in case-control studies if exposure information is obtained retrospectively.

9.6.3 Extension to multiplicative error

Regression calibration can also be extended for use when the error is multiplicative, that is, of the form

$$X_m = X_t \omega, \tag{9.37}$$

where ω is the error. Taking logs of both sides turns this into the classical error model on a log transformed scale. Suppose we assume that $\log X_t$ and $\log X_m$ have a joint normal distribution. Under this assumption the regression of $\log X_t$ on $\log X_m$ is linear; we write

$$\log X_t = \gamma_0 + \gamma_{t|m} \log X_m + \tau \tag{9.38}$$

and denote the variance of the residuals by σ_τ^2. It can be shown that

$$\mu_{X_t|X_m} = e^{\gamma_0 + \sigma_\tau^2} X_m^{\gamma_{t|m}}, \tag{9.39}$$

$$\sigma_{X_t|X_m}^2 = e^{\gamma_0 + \sigma_\tau^2}(e^{\sigma_\tau^2} - 1)X_m^{2\gamma_{t|m}}. \tag{9.40}$$

These may be substituted into the expression for $E(Y|X_m)$ given in (9.34) to obtain an approximate expression for $E(Y|X_m)$ in the multiplicative error setting:

$$E(Y|X_m) \approx L_1 \left\{ \frac{\alpha^* + \beta_t e^{\gamma_0 + \sigma_\tau^2} X_m^{\gamma_{t|m}}}{(1 + k^2 \beta_t^2 e^{\gamma_0 + \sigma_\tau^2}(e^{\sigma_\tau^2} - 1)X_m^{2\gamma_{t|m}})^{1/2}} \right\}. \tag{9.41}$$

If the denominator of this is close to 1 then it follows that β_t can be estimated by replacing X_t in the logistic regression by

$$e^{\gamma_0 + \sigma_\tau^2/2} X_m^{\gamma_{t|m}}.$$

From this it can be seen that multiplicative error turns a true linear relationship on the logistic scale into a non-linear relationship using the mismeasured exposure.

9.6.4 Incorporation of a simple form of differential error

The method of regression calibration relies on the assumption that the error in X_m is non-differential. In this section we show that, under assumptions about the distribution of the true exposure X_t, corrections for one type of differential error can be made using an approach not dissimilar to regression calibration. We focus here on an unmatched study, though the explanations below can be extended to studies in which cases and controls are sampled within strata. Suppose now that among controls the true exposure X_t arises from a normal distribution with mean μ_{X_t} and variance κ^2 and that the corresponding distribution of X_t among cases is normal with mean $\mu_{X_t} + \Psi$ and variance κ^2.

We assume the following error models for X_m among cases and controls:

$$X_m = X_t + \epsilon_y, \quad y = 0, 1, \tag{9.42}$$

where the errors ϵ_y are normally distributed with mean $\gamma + y\delta$ ($y = 0, 1$) and variance τ^2. Differential error is incorporated via the term δ, and γ represents a constant mean shift. For simplicity, suppose that each individual in the case-control study has a single measurement of X_m. A naive analysis using the mismeasured exposure is based on the model

$$\Pr^D(Y = 1 | X_m = x_m) = \frac{\Pr^D(X_m = x_m | Y = 1) \Pr^D(Y = 1)}{\Pr^D(X_m = x_m)}. \tag{9.43}$$

Using the distributional assumptions outlined above, it can be shown that

$$\Pr^D(Y = 1 | X_m = x_m) = L_1(\alpha^{**} + \beta^* x_m), \tag{9.44}$$

where

$$\alpha^{**} = \log \frac{\Pr^D(Y = 1)}{\Pr^D(Y = 0)} - \frac{(\Psi + \delta)^2}{2(\kappa^2 + \tau^2)} - \frac{(\mu_{X_t} + \gamma)(\Psi + \delta)}{(\kappa^2 + \tau^2)},$$
$$\beta^* = \frac{\Psi + \delta}{\kappa^2 + \tau^2}.$$

It follows that under the above assumptions the naive model, $\Pr^D(Y = 1 | X_m = x_m)$, is of logistic form. When there is no measurement error, $\delta = \gamma = \tau^2 = 0$ and (9.44) reduces to

$$\Pr^D(Y = 1 | X_m = x_m) = L_1(\alpha^* + \beta_t x_m), \tag{9.45}$$

where β_t is the log odds ratio parameter of interest and is given by $\beta_t = \Psi / \kappa^2$. It follows that, under the assumptions and error model above, the corrected form β_t is given by

$$\beta_t = \frac{\beta^*(\kappa^2 + \tau^2) - \delta}{\kappa^2}. \tag{9.46}$$

The parameters required to obtain this correction can be estimated in a validation study within the case-control sample in which either X_t can be observed or one or more repeated measures of X_m is or are available.

The form of differential error considered above is simple, being in the form of a mean shift in the measurements. The results do not extend to more complex forms of differential error, such as those for which the error variability differs in the cases and the controls.

9.6.5 Methods allowing for differential error

In this section we outline a further two, fairly recently proposed, methods for measurement error correction that can accommodate differential error, namely multiple imputation and moment reconstruction. Both methods allow for differential error because they are designed to construct imputed values of the true exposure, conditionally on all relevant aspects of the data including the outcomes. The imputed values are used directly in the regression analysis of interest.

We begin by considering multiple imputation, which originated as a method for handling missing data and is now widely used in that context. More recently an application of this approach in measurement-error problems has been described in which the true exposure X_t is treated as missing data. Here, the key idea in multiple imputation is that true exposure measurements are imputed by drawing a random value from the distribution of the true exposure conditional on all observed values. To account for the uncertainty inherent in a single imputed value, under multiple imputation several imputations are created for each missing data point, thus producing several versions of the complete data. The main analysis model, in this case a logistic regression, is then repeated within each imputed data set and the resulting parameter estimates are pooled.

Here, the true exposure X_t is imputed for all individuals by drawing values from the distribution $X_t|X_m = x_m, Y = y$, where X_m, Y are fully observed. This conditional distribution is assumed known. The imputed values are then used in the main analysis model and the resulting estimates are pooled across imputed data sets. It has been shown that it is important that the outcome Y is included in the imputation model for X_t. The distribution of the true exposure X_t conditionally on X_m and Y is, in general, a non-standard distribution. An approximate imputation model is therefore often used. A commonly used such model is of the form

$$X_t = \theta_0 + \theta_1 X_m + \theta_2 Y + \zeta, \tag{9.47}$$

where the residuals ϵ are normally distributed with mean 0 and variance σ_ϵ^2. For a given individual the imputed measurement is given by

$$X_t^* = E(X_t | X_m = x_m, Y = y) + \zeta^*,$$

where ζ^* is a random draw from the distribution of the residuals. The procedure is repeated to give M imputed data sets. The model for the exposure-outcome association is fitted within each imputed data set and the resulting estimates combined. The combining of estimates was outlined in Section 7.11. In any application of multiple imputation, correct specification of the imputation model is important. When the outcome involves a time component, as in the studies described in Chapters 7 and 8, this should be incorporated into the imputation model. The method of multiple imputation also accommodates binary or categorical exposures, in which case the imputation model may be a logistic regression.

In a case-control study the imputation model (9.47) may be fitted using a validation study in which X_t is observed within a random subset of the cases and controls. Where X_t cannot be observed the imputation model can be fitted if replicate values of the measured exposure X_m are available for at least some individuals, under some assumptions about the error model. We do not give the details of this here.

Moment reconstruction is another imputation-based approach to measurement-error correction that accommodates differential error. The idea behind this method is to find values, which we denote X_{MR}, such that the first two *joint* moments of X_{MR} with Y are the same as the first two joint moments of X_t with Y: $\sum X_t Y = \sum X_{MR} Y, \sum X_t^2 Y = \sum X_{MR}^2 Y$.

The moment-reconstructed values are obtained as a function of the mismeasured exposure X_m and the outcome Y. It can be shown that X_{MR} can be obtained using the formula

$$X_{MR}(X_m, Y) = E(X_m | Y) + \{X_m - E(X_m | Y)\} \sqrt{\frac{\text{var}(X_t | D)}{\text{var}(X_m | Y)}}. \tag{9.48}$$

When $X_t | Y$ and $X_m | Y$ have a jointly normal distribution given Y, it follows that the joint distribution of X_{MR} and Y is the same as that of X_t and Y. This is in contrast with the regression calibration values, which agree with the true exposure values only in expectation.

The elements required to find the moment-reconstructed values X_{MR} can be estimated in a validation study in which X_t is observed. Under the classical error model, potentially with different error variances in cases and controls, the moment-reconstructed values can be obtained using repeated exposure measurements X_m.

It is worth noting that if differential error is assumed when in fact the error is non-differential, a loss of efficiency in the error correction will result. If the error is non-differential then regression calibration is likely to be the more efficient correction approach.

Like regression calibration, the methods outlined in this section can be extended to accommodate adjustments for additional variables W in the exposure-outcome model.

9.6.6 A full likelihood approach

Corrections for measurement error in continuous exposures can also be performed by writing down the full likelihood of the observed data and incorporating the assumed error model. In Section 9.2.4 a likelihood-based approach was outlined for the case of a binary exposure measured with error. Here we briefly outline an extension to continuous exposures.

We focus on a case-control study in which the measured exposure X_m is observed for all individuals and the true exposure X_t is observed in a validation subsample comprising n_{1V} cases and n_{0V} controls ($n_{1V} + n_{0V} = n_V$). The numbers of cases and controls in the remainder of the case-control sample are denoted respectively n_{1R} and n_{0R} ($n_{1R} + n_{0R} = n_R$). The full likelihood for the observed data is proportional to

$$\prod_{i \in V} f^V_{X_t X_m | Y}(x_{ti}, x_{mi} | y_i) \prod_{i \in R} f^R_{X_m | Y}(x_{mi} | y_i), \tag{9.49}$$

where the superscript V denotes conditioning on being in the validation sample and R denotes conditioning on being in the remainder of the study sample. We note the connection with the two-stage sampling designs discussed in Chapter 6 and the full likelihoods outlined there.

The contributions from individuals in the validation sample can be written as

$$f^V_{X_t X_m | Y}(x_t, x_m | y) = f^V_{X_m | X_t, y}(x_m | x_t, y) L_y(\alpha^*_V + \beta_t x_t) f^V_{X_t}(x_t) n_V / n_{yV}. \tag{9.50}$$

The contributions from individuals in the remainder of the study sample are found by integrating over the missing true values:

$$f^R_{X_m|Y}(x_m|y) = \int f^R_{X_m|X_tY}(x_m|x_t, y)L_y(\alpha^*_R + \beta_t x_t)f^R_{X_t(x_t}n_R/n_{yR}\,dx_t. \quad (9.51)$$

A parametric distribution must be specified for $f_{X_m|X_tY}$. To evaluate the integral in (9.51) a parametric model for the marginal distribution of X_t could also be specified; however, this specification may be undesirable, and methods have been proposed that treat the integral non-parametrically.

9.6.7 Extensions to matched case-control studies

The correction methods described above can all be applied in matched case-control studies, though they may not yet have been much used in this context. Care should be taken over how the matching is incorporated in any covariate conditioning. For example, in regression calibration it seems appropriate for the matching variables to be included as predictors in the model for X_t, though how important this is must depend on how strongly the matching variables are associated with the exposure and outcome. Likewise, in multiple imputation and moment reconstruction, conditioning on the matching variables in the imputation models is appropriate if possible. The full likelihood does not appear best suited to dealing with error in matched studies, unless the matching is broken and an unconditional analysis is used.

Notes

Section 9.1. For general discussions of the various kinds of measurement error in regression-like situations see, for example, Carroll *et al.* (2006), who gave a comprehensive account, Buonaccorsi (2010), Gustafson (2003) and Reeves *et al.* (1998).

Section 9.2.2. For simple correction methods for the misclassification of a binary exposure see Barron (1977), Greenland and Kleinbaum (1983) and Morrissey and Spiegelman (1999). Hui and Walter (1980) described methods for estimating error rates, including those in a situation using only replicate measures. In general, at least three observations of the misclassified exposure are required to estimate misclassification probabilities in this approach. For another approach using only replicates plus perfectly measured covariates see Kosinski and Flanders (1999). For misclassification in a survey context, see Kuha and Skinner (1997), Kuha *et al.* (1998) and, in a more epidemiological setting, Duffy *et al.* (2004). Bayesian methods for misclassification corrections have been described by Prescott and Garthwaite (2005) and Chu *et al.* (2010).

Rice and Holmans (2003) focused specifically on allowing misclassification of the genotype in unmatched case-control studies of associations between genotype and disease, where the exposure has three levels (for example aa, aA, AA). For studies of gene-environment interactions, Zhang *et al.* (2008) described methods for obtaining corrected estimates when both the genetic and the environmental exposures may be subject to misclassification.

Section 9.2.4. The likelihood-based approach was described by Carroll *et al.* (1993).

Section 9.3. See Carroll *et al.* (2006, Chapter 15) for an overview of the effects of error in response variables, including the misclassification of binary outcomes in logistic regression. Methods for correcting for the effects of binary outcome misclassification in logistic regression were also outlined by Carroll *et al.* (2006, Chapter 15) for different scenarios.

Section 9.4. Methods for correcting for exposure misclassification in pair-matched case-control studies were developed by Greenland (1982), Greenland and Kleinbaum (1983) and Greenland (1989). Variance estimates for corrected odds ratio in the matched-pair situation were discussed by Greenland (1989). An extension to the case of matched case-control studies with variable numbers of controls per case was given by Duffy *et al.* (2003). The likelihood-based approach was described by Rice (2003).

Section 9.5. Berkson error is named after Berkson (1950). See Reeves *et al.* (1998) for comments on mixtures of classical and Berkson error and for an an argument showing that Berkson error can result in the attenuation of estimates under logistic regression. Heid *et al.* (2004) gave a detailed discussion of classical and Berkson error in the context of studies of radon exposure. Some special problems of measurement error in nutritional epidemiology, as considered in Example 9.4, are discussed in Willett (1998). See, for example, Kipnis *et al.* (2001) for a discussion of error models for dietary measurements.

Section 9.6.1. Many approaches to measurement error correction have emerged and we outline only a few of them here. Regression calibration is probably the most commonly used. The imputation-based approaches have attractive properties and are relatively simple to implement, though being fairly new have not yet been used much in practice. Thürigen *et al.* (2000) gave a useful review of methods for error correction in case-control studies when the true exposure is observed in a validation subsample. Other methods not covered here include simulation extrapolation, or SIMEX, which has also been referred to as 'shaking the data'. Under this method the investigator adds additional error to an already error-prone exposure measurement in a systematic way. The effect on the resulting estimate of interest, in this case the log odds ratio, is then modelled as a function of the degree of error and this model is used to extrapolate back to the situation of no measurement error. There may be considerable uncertainty attached to the extrapolation. The SIMEX method was described in Cook and Stefanski (1994); see also Carroll *et al.* (2006, Chapter 5).

Further methods not mentioned in detail include conditional score methods; see Carroll *et al.* (2006, Chapter 7) and McShane *et al.* (2001) for their use in matched case-control studies. Methods using estimating equations were described by Carroll *et al.* (1995). A number of authors have considered the use of Bayesian methods in measurement-error correction; see, for example, Müller and Roeder (1997) and Gustafson *et al.* (2002). In some studies there may be multiple exposures of interest, more than one of which is measured with error. While classical non-differential measurement error in a single exposure will result in an attenuation of the main effect, this does not extend, however, to the multivariable case. Multivariable classical measurement error can have the result that exposure-outcome associations are biased either away from or towards the null. Regression calibration has been extended to this situation.

Section 9.6.2. The use of regression calibration for logistic models was first described by Rosner *et al.* (1989, 1992). See also Carroll *et al.* (2006, Chapter 4), Reeves *et al.* (1998) and Kuha (2006). The original reference for the probit approximation to the logistic is Cox (1966); see also Carroll *et al.* (1984) and Carroll *et al.* (2006, p. 91). The use of regression calibration in proportional hazards regression was considered by Prentice (1982), Hughes (1993) and Spiegelman and Rosner (1997) and the simple approximation in which $E(X_t | X_m)$ is used in place of X_t has been found to work well under many conditions. Regression calibration has been extended for models involving non-linear functions of the exposure; see Carroll *et al.* (2006, p. 94) and Cheng and Schneeweiss (1998). The effects of classical measurement error in models involving non-linear exposure terms were illustrated by Keogh *et al.* (2012a). The extension of regression calibration to a multivariable situation was described by Rosner *et al.* (1990).

Section 9.6.3. Regression calibration for multiplicative error was outlined by Reeves *et al.* (1998) and Carroll *et al.* (2006, Section 4.5).

Section 9.6.4. The method of regression calibration for the simple form of differential error described in this section was proposed by Armstrong *et al.* (1989), who outlined the method in general terms for a case-control study with sampling within strata and for multiple exposures measured with error. They illustrated the method in a study of the association between dietary intake and colon cancer.

Section 9.6.5. We refer to the notes of Chapter 7 for some general literature on multiple imputation. Cole *et al.* (2006) and Freedman *et al.* (2008) proposed the use of multiple imputation to correct for measurement error in continuous exposures. Moment reconstruction was proposed by Freedman *et al.* (2004). Freedman *et al.* (2008) compared multiple imputation, moment reconstruction and regression calibration in simulation studies, including a case-control setting where there is differential error. Thomas *et al.* (2011) proposed an extension to moment reconstruction that they call 'moment adjusted imputation'. This allows for the construction of values which are such that, instead of matching the first two moments of X_t with Y, the first m moments are matched ($m \geq 2$). This method may be useful, for example when $\Pr(Y = 1|X_t)$ is non-linear in X_t on the logistic scale or when X_t is non-normal.

Section 9.6.6. The full likelihood method outlined in this section comes from Carroll *et al.* (1993), who gave detailed formulae for the estimating equations and standard errors. Related approaches were described by Roeder *et al.* (1996). These included an approach in which the marginal distribution of X is treated non-parametrically by using the expectation-maximization (EM) algorithm, which allows an extension of the method outlined in this section to the situation in which X_t is not observed in a subsample but external information or replicate measures of X_m are available. Another related method was described by Fearn *et al.* (2008). Case-control studies with a validation sub-sample may be considered as a two-stage design (see Schill *et al.* (1993), for example), and there is a link with the methods described for two-stage studies in Chapter 6.

Section 9.6.7. Methods for dealing with error in matched case-control studies have received less attention. See McShane *et al.* (2001) for the use of regression calibration in matched case-control studies. See also Reeves *et al.* (1998) and Fearn *et al.* (2008). A likelihood-based correction method for matched case-control studies was described by Guolo and Brazzale (2008), who also made comparisons with other approaches.

10

Synthesis of studies

- The accumulation and possible combination of evidence from different studies, as contrasted with an emphasis on obtaining individually secure studies, is crucial to producing convincing evidence of generality of application and interpretation.
- Methods for combining estimates across studies include those that can be applied when the full data from individual studies are available and those that can be applied when only summary statistics are available. They make allowance for heterogeneity of the effect of interest. The methods are general and not specific to case-control studies.
- Some special considerations may be required for the combining of results from case-control studies and full cohort studies or from matched and unmatched case-control studies.

10.1 Preliminaries

The emphasis throughout this book has been on analysis and design aimed to produce individually secure studies. Yet in many fields it is the accumulation of evidence of various kinds and from various studies that is crucial, sometimes to achieve appropriate precision and, often of even more importance, to produce convincing evidence of generality of application and interpretation. We now consider briefly some issues involved, although most of the discussion is not particularly specific to case-control studies. The term *meta-analysis* is often used in this context. The distinctive issues are, however, not those of analysis but more those of forming appropriate criteria for the inclusion of data and of assessing how meaningfully comparable different studies really are. Important issues of the interpretation of apparent conflicts of information may be best addressed by the traditional method of descriptive review.

We suppose that either the full data, or perhaps more commonly only the results of summary analyses, are available for m independent studies of the effect of the same exposure variable or variables on the same outcome measure. The first and crucial consideration concerns the appropriateness of the studies chosen for inclusion. Aspects of this include:

- the genuine comparability of exposure and outcome criteria, including the recording of details that may possibly be used for adjusting for relatively minor lack of comparability;
- the assurance of quality;
- checks for the absence of selection bias, in particular that arising from publication bias in favour of studies reporting some positive effect as contrasted with those with an apparently null conclusion.

Bias of the last kind is to be suspected if, for example, small studies appear to show a larger effect of exposure than apparently comparable large studies.

The synthesis of observational studies, and of case-control studies in particular, in some ways raises more general challenges than the synthesis of experimental investigations, for example randomized clinical trials. This is so, in particular because of concern in observational studies about observed and unobserved confounders. Assessment of these in different studies, the environment for which is incompletely known, may be difficult. There may also be concerns connected with a lack of exact comparability of definition of apparently similar parameters arising in different types of model.

10.2 Some large-sample methods

Suppose that there are m independent studies and that there is available from the kth study an estimate $\hat{\psi}_k$ of the relevant log odds ratio and its large-sample variance v_k, all the studies being large enough for asymptotic methods of analysis to be appropriate. These will often be estimated after adjustment for lack of balance in intrinsic features between exposed and unexposed groups. Assuming provisionally that the separate studies do indeed estimate the same value ψ, the method of weighted least squares may be used for combining them. The methods used are essentially those outlined in Chapter 2 in connection with the combining of estimates across strata. To avoid the use of special formulae, note that the derived observations $\hat{\psi}_k/\sqrt{v_k}$ have constant unit variance and obey the large-sample linear model

$$E \begin{pmatrix} \hat{\psi}_1/\sqrt{v_1} \\ \cdot \\ \cdot \\ \cdot \\ \hat{\psi}_m/\sqrt{v_m} \end{pmatrix} = \begin{pmatrix} 1/\sqrt{v_1} \\ \cdot \\ \cdot \\ \cdot \\ 1/\sqrt{v_m} \end{pmatrix} \psi. \tag{10.1}$$

The method of ordinary least squares may now be used with (10.1), to give

$$\hat{\psi} = \frac{\Sigma \hat{\psi}_k / v_k}{\Sigma 1 / v_k},$$ (10.2)

where

$$\text{var}(\hat{\psi}) = \frac{1}{\Sigma 1 / v_k}.$$ (10.3)

The residual sum of squares of the derived observations is

$$\Sigma \{ (\hat{\psi}_k - \hat{\psi}) / \sqrt{v_k} \}^2 = \Sigma \hat{\psi}_k^2 / v_k - \hat{\psi}^2 \Sigma 1 / v_k.$$ (10.4)

This is a residual sum of squares corresponding to a unit error variance, and consistency with this may therefore be tested by comparison with the chi-squared distribution with $m - 1$ degrees of freedom. Too large a value of the test statistic signals a need to reformulate the situation to take account more realistically of the additional variation detected in the $\hat{\psi}_k$. Methods which allow for heterogeneity between studies are discussed in a later section.

Sometimes one or more explanatory variables w_k associated with each whole study may be available. For example, if each study was done in a very different geographical area then w_k might be a summary measure of the climate in the area. The relevance of such a variable may be tested by the linear model

$$E \begin{pmatrix} \hat{\psi}_1 / \sqrt{v_1} \\ \cdot \\ \cdot \\ \hat{\psi}_m / \sqrt{v_m} \end{pmatrix} = \begin{pmatrix} 1/\sqrt{v_1} & w_1/\sqrt{v_1} \\ \cdot & \cdot \\ \cdot & \cdot \\ 1/\sqrt{v_m} & w_m/\sqrt{v_m} \end{pmatrix} \begin{pmatrix} \psi \\ \gamma \end{pmatrix}.$$ (10.5)

This leads to the estimate

$$\hat{\gamma} = \frac{\Sigma (w_k - \tilde{w})(\hat{\psi}_k - \hat{\psi}) / \sqrt{v_k}}{(\Sigma 1 / \sqrt{v_k}) \Sigma (w_k - \tilde{w})^2 / v_k}$$ (10.6)

with variance

$$\frac{1}{(\Sigma 1 / v_k) \Sigma (w_k - \tilde{w})^2 / v_k}.$$ (10.7)

Here

$$\tilde{w} = \frac{\Sigma w_k / v_k}{\Sigma 1 / v_k}.$$ (10.8)

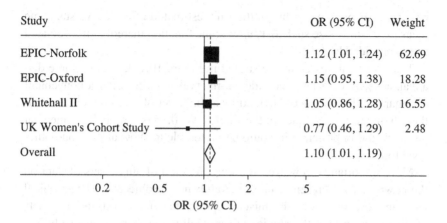

Study	OR (95% CI)	Weight
EPIC-Norfolk	1.12 (1.01, 1.24)	62.69
EPIC-Oxford	1.15 (0.95, 1.38)	18.28
Whitehall II	1.05 (0.86, 1.28)	16.55
UK Women's Cohort Study	0.77 (0.46, 1.29)	2.48
Overall	1.10 (1.01, 1.19)	

Figure 10.1 Adjusted odds ratio estimates for breast cancer per 10 g/day increase in alcohol intake, showing separate estimates within four case-control studies and a combined estimate.

10.3 Graphical representation

Associated with the weighted least squares analyses of the previous section are simple graphical procedures. Probably the most widely used of these is illustrated in Figure 10.1. Each study is summarized by the odds ratio for the effect of a binary exposure together with 0.95-level confidence limits calculated first for the log odds ratio and then exponentiated. At the bottom is the summarizing estimate (10.2) and its associated confidence interval.

Example 10.1: Synthesis of studies Keogh *et al.* (2012b) combined estimates of the association between alcohol intake and breast cancer from four case-control studies, all of which were part of the UK Dietary Cohort Consortium and were nested within UK cohorts for which prospective information in dietary intake was obtained (EPIC-Norfolk, EPIC-Oxford, Whitehall II and the UK Women's Cohort Study). Combined across the four studies there were 656 breast cancer cases and 1905 matched controls. A combined measure of alcohol intake was obtained from each study and the authors reported the odds ratio for a 10 grams per day increase in alcohol intake. The odds ratios were adjusted for a number of potential confounders, which we do not list here. The estimates within studies and a combined estimate are summarized in Figure 10.1.

One possible source of bias arises from selective reporting of small studies showing an apparent exposure effect combined with the under-reporting of small studies with no such effect. As noted above, this may be tested to some extent by plotting the estimated effect $\hat{\psi}_k$ versus some measure of study size such as $1/v_k$ or some function thereof. An associated test is provided by (10.4).

The graphical procedure illustrated in Figure 10.1 does have some disadvantages. A 0.95-level confidence interval is quite often a convenient summary of the evidence provided by a single set of data on their own, although of course the choice of the single level 0.95 is essentially somewhat arbitrary and in principle it would be preferable to show at least one other level too.

More importantly, however, the interpretation of single sets of data on their own is, in the present context, certainly not the objective. The graphical form invites, moreover, the misconception that if intervals for two sets of data overlap then the studies are mutually consistent. Conversely, the presence of a study whose interval does not overlap the summarizing value corresponding to $\hat{\psi}$ is not necessarily evidence of heterogeneity. Nor if the individual confidence intervals overlap are the studies necessarily best treated as homogenous.

These objections would be overcome to some extent by showing not 0.95-level confidence intervals but plus-or-minus-one standard error bands. In particular this would lead slightly more easily to the comparison of two or more individual studies and would not so strongly invite the misconceptions just mentioned, nor the impression that all individual estimates lie within the relevant band.

There is, however, a much more critical issue connected with such analyses, which we now discuss.

10.4 Study heterogeneity

10.4.1 Preliminaries

The supposition underlying the above discussion is that the value of ψ is identical in all studies. If in fact estimating a fundamental physical constant were the objective then the assumption of constant ψ would need checking as a precaution against, say, additional systematic experimental error, but would be a natural basis for analysis. For the kinds of application for which case-control studies are likely to be used, the assumption of homogeneity, even as a rough approximation, is much less plausible. There are a number of reasons for heterogeneity, including the following.

- The mix of subjects involved is likely to be different in different studies, and adjustment for only some of the relevant features may be feasible.

- In connection with the previous point the initial estimates $\hat{\psi}_k$ may have been be adjusted for different intrinsic features in different studies.
- Exposure definitions may not be precisely comparable.
- Procedures for selecting controls, and hence for defining the basis for comparison, may be different.

The conventional procedure is to examine for general heterogeneity by the overall chi-squared statistic (10.4) and to assume homogeneity in the absence of a significant departure. Such a global test is, however, insensitive to specific focused types of departure and, wherever feasible, more specific tests should be applied for the kinds of heterogeneity listed above. In particular, one possibility would be to take the potentially most important intrinsic variables in turn, to classify the studies according to whether adjustment was made for a given variable and to test the agreement of the ψ values for the two sets of estimates. Analyses may also be made for dependence on study size and possibly on study quality, if the latter can reasonably be assessed.

The consequences of non-homogeneity for an analysis that assumes homogeneity are essentially twofold. One is that the precision of the overall assessment is exaggerated; a confidence interval for $\hat{\psi}$ based on (10.3) can be unrealistically narrow when a number of studies are combined.

A second, and in some ways more important, aspect is that if one study has a much smaller variance than the others, usually because it is much bigger, the results of that study have a very strong and undue influence on ψ. We return to this aspect below.

10.4.2 Representation of heterogeneity

If heterogeneity is detected among the estimates $\hat{\psi}_k$ then, if at all possible, one or possibly several rational explanations should be found. If none is available and it is decided to proceed quantitatively the usual representation is

$$\psi_k = \psi_{\cdot} + \eta_k, \tag{10.9}$$

where η_k is a random variable of zero mean and variance σ_η^2 that is possibly normally distributed. That is,

$$\hat{\psi}_k = \psi_{\cdot} + \eta_k + \epsilon_k, \tag{10.10}$$

where ϵ_k is the internal error in the estimation of zero mean and variance v_k.

Before studying the implications of this for analysis, consider the interpretation of (10.9) and especially of the parameter $\psi.$. This is sometimes interpreted as referring to an average effect over a population of studies, of which the studies under analysis are regarded as a random sample. Such a population is in many contexts so artificial as to be virtually without useful meaning. A possibly helpful interpretation is the following. The process or processes generating variation in the log odds ratio are probably complicated and unknown and may be represented roughly as a stochastic system. The parameter $\psi.$ represents the result of averaging over that process.

If the variance σ_η^2 were to be treated as known then the optimal estimate of $\psi.$ would be the weighted mean

$$\frac{\Sigma \hat{\psi}_k/(\sigma_\eta^2 + v_k)}{\Sigma 1/(\sigma_\eta^2 + v_k)}, \tag{10.11}$$

with variance

$$\frac{1}{\Sigma 1/(\sigma_\eta^2 + v_k)}. \tag{10.12}$$

The most important aspect of this is often that it places less emphasis on very large studies having relatively small values of v_k, thus moving the estimate towards an unweighted mean of the separate estimates.

The formal procedure of analysis, assuming approximate normality of the distributions involved, is to apply the method of maximum likelihood to the log likelihood

$$l(\psi., \sigma_\eta^2) = -\sum \frac{\log(\sigma_\eta^2 + v_k)}{2} - \sum \frac{(\hat{\psi}_k - \psi.)^2}{2(\sigma_\eta^2 + v_k)}. \tag{10.13}$$

Effective estimation of the component of variance σ_η^2 requires at least a modest number of studies, say 15 or more.

Note that at every stage of an iterative procedure the current estimate of $\psi.$ is the weighted mean of the $\hat{\psi}_k$. Moreover, the off-diagonal element of the Fisher information matrix is zero:

$$E\left\{-\frac{\partial^2 l(\psi., \sigma_\eta^2)}{\partial \psi. \partial \sigma_\eta^2}\right\} = 0. \tag{10.14}$$

This implies that, asymptotically, the sampling errors in the estimates of the two parameters are independent and, in some ways more importantly, that the estimate of $\psi.$ is relatively insensitive to estimation errors in the variance component.

If the studies are, so far as can be ascertained, of equal quality and the additional variation in the estimates represented by σ_η^2 is real rather than an artefact then undue reliance on a single study, even if large, is indeed unwise. However, if the quality of the large study is high and some of or all the extra variation is artificial then the emphasis on large studies is justified.

In practice we should perform a sensitivity analysis with respect to σ_η^2.

We now discuss briefly what might be done if the full data from the individual studies is available for integrated analysis.

10.4.3 Using the originating data

As noted above, in some situations the combined analysis may have to be based only on summary analyses of the primary studies. In others there may be access to the originating data, often called *individual patient data* in a medical context, opening up the possibility of re-analysis so far as feasible by common methods. When the full data are available there is thus an important initial choice to be made between two possible approaches:

- a one-step approach, that is, a single composite analysis of all available studies;
- a two-step approach in which we proceed first to individual studies and second to a combination of estimates using methods such as those outlined above.

While a single overall interpretation is in general much to be preferred to fragmentary conclusions, if the individual studies are of viable size then it may be preferable to start with separate analyses and examine carefully the extent to which the conclusions reinforce one another. The simplest one-step approach may incorporate fixed study effects, but it makes an assumption of homogeneity of the effect of the exposure across studies. The two-step approach can allow for heterogeneity between studies by the incorporation of random effects as in (10.9).

When the study effects are homogeneous, the one-step and two-step methods coincide in the simple situation of a linear model with fixed exposure and study effects. However, in most situations, the results obtained using one-step and two-step approaches will differ even if only slightly. When there is heterogeneity of the exposure effect across studies and a random effects model is appropriate, the two-step method gives unbiased estimates of the average exposure effect and its standard error when the individual studies are large, while the simple one-step method, which assumes a fixed exposure effect, gives biased estimates with underestimated standard error.

When individual patient data is used, a primary source of practical difficulty may be the non-availability, in some studies, of important intrinsic variables; this brings difficulties in performing a one-step analysis. However, the availability of individual patient data can bring many advantages, including the ability to adjust for common variables across studies, where they have been measured, and possibly the ability to use a common definition of exposure across studies, where it may have differed in the original analyses of single studies.

Example 10.1 continued   In the study of alcohol intake and breast cancer described in Example 10.1, in fact individual data was available for each of the four studies. This enabled the authors to use a common measure of alcohol intake across the studies. It also enabled adjustment for common variables in the analysis and for these to be defined, as far as possible, in the same way; for example, it was desirable to adjust for a measure of physical activity, but this had been measured in different ways across the studies. A common four-category variable was derived using the different information available in the different studies. Within studies there was some missing data on several adjustment variables. Multiple imputation was used to impute missing data within each study and estimates were combined across imputed data sets to obtain a single estimate in each study. These were combined using the two-step methods described above, assuming homogeneity of the effect across studies. The value found for the statistic in (10.4) was 2.27. Comparing this with a chi-squared distribution with three degrees of freedom shows no evidence of heterogeneity between the studies. However, this investigation involved only four studies.

10.5 An exact analysis

Our earlier discussion in this chapter concentrated on the method of weighted least squares as the primary method of analysis, partly because of wide familiarity with at least the unweighted least squares method and the availability of flexible computer packages for its implementation, but more importantly because of its flexibility and the availability of procedures for model checking and supplementation. Nevertheless, in some simple situations more direct methods are available and are especially appropriate when small frequencies are involved. We now outline this approach.

As shown in Section 2.3, inference about the log odds ratio when the exposure is binary is summarized in the generalized hypergeometric distribution

$$\frac{\binom{n_1}{r_1}\binom{n_0}{r_0}e^{r_1\psi}}{\Sigma\binom{n_1}{s}\binom{n_0}{r_. - s}e^{s\psi}}. \tag{10.15}$$

Here r_1 and r_0 are respectively the numbers of exposed individuals among n_1 cases and n_0 controls.

We consider m studies in which the different probabilities in the associated 2×2 tables are distinct, but in which ψ is assumed constant. Denote the number of cases in study k by n_{1k}, of which r_{1k} are exposed. A corresponding notation is used for controls. Then sufficiency arguments show that the distribution for inference about ψ is the conditional distribution of $R_{1.} = \Sigma R_{1k}$ given the margins in the individual tables. The required distribution is the convolution of m generalized hypergeometric distributions (10.15). In general this can be used directly only by numerical computation.

In the special case where we are examining the consistency with $\psi = 0$, the individual component distributions are hypergeometric. The moments of this distribution can be calculated and it follows that

$$E(R_{1.}; \psi = 0) = \sum \frac{n_{1k}r_{.k}}{n_{1k} + n_{0k}} \tag{10.16}$$

and

$$\text{var}(R_{1.}; \psi = 0) = \frac{n_{1k}n_{0k}r_{.k}(n_{.k} - r_{.k})}{n_{.k}(n_{.k} - 1)}. \tag{10.17}$$

Higher moments could be calculated, but in fact an appeal to the central limit theorem for $R_{1.}$ suggests that a normal approximation with continuity correction will almost always be adequate. Various approximations for computing confidence limits are possible but the most secure method is to compute and plot the conditional log likelihood function $l_c(\psi)$ as a function of ψ and to take all values within an appropriate distance of the maximum; an approximate $1 - 2\alpha$ confidence region is given by

$$\{\psi : l_c(\hat{\psi}) - l_c(\psi) \leq k_\alpha^{*2}/2\}, \tag{10.18}$$

where k_α^* is the upper α point of the standard normal distribution. A number of alternative methods based essentially on local expansions near $\hat{\psi}$ are available. If, however, more detailed analyses are of interest, including the assessment of dependence on study features, then the methods based on least squares will be more flexible and are likely to be adequate unless some quite small cell frequencies play an important role.

10.6 Combining studies of different type

In some ways the most delicate issues involving the combining of observations from different investigations arise when the studies are of different types, for example when matched and unmatched studies or case-control studies are taken together with broadly comparable prospective studies. Some quantitatively meaningful basis for comparison must usually be assumed, although with a large number of studies of very different kinds the apparent direction of an effect might be a basis for a rather crude overall assessment. In other contexts, as for example when we are combining the results of laboratory measurements with those from clinical trials, the traditional methods of systematic qualitative review have to be used.

The primary analysis of an unmatched case-control study by some form of logistic regression is by construction designed to establish comparability with a notional prospective study, so that in principle the combining of results from prospective and retrospective studies should raise no special issues. In practice, especially if there are a number of studies of each type, it will be wise to check for possible systematic differences between the conclusions from different types before embarking on a synthesis.

The formulations standard for unmatched and for matched studies appear similar. Typically for, say, a binary exposure there is a parameter β characterizing the differences in the log odds for cases and controls. In the simplest situation this has a direct interpretation for unmatched studies whereas for pair-matched studies the parameter β is defined conditionally on parameters specifying the inter-pair systematic differences. Because of the nonlinearity of the logistic function, such a parameter is typically greater than would be estimated from an unmatched study, although the difference may be small.

Notes

Section 10.1. One of the first discussions of the combining of studies was by Yates and Cochran (1938). Although their account focused on agricultural field trials, many of their points are of broad relevance. Hedges and Olkin (1985) focused on the combination of independent tests of significance; here we have emphasized the estimation of effects. For a fine example of the synthesis of information from studies of different types, see the influential review by

Cornfield *et al.* (1959) of evidence concerning tobacco smoking and lung cancer. A recent review of methods for pooling the results of epidemiologic studies is given in Smith-Warner *et al.* (2006), with reference to the pooling of results from studies of associations between diet and cancer. There is a large recent literature on statistical methods in meta-analysis. Sutton and Higgins (2008) gave a review of recent developments and noted the available software. Borenstein *et al.* (2010) gave an introduction to fixed-effects and random-effects meta-analysis.

Section 10.2. The use of the statistic in (10.4) to assess heterogeneity was described by Cochran (1954).

Section 10.3. Anzures-Cabrera and Higgins (2010) discussed graphical displays for the synthesis of studies.

Section 10.4.2. An often quoted paper in this area is that of DerSimonian and Laird (1986), who outlined the use of random effects to incorporate heterogeneity in estimates from different studies. Thompson and Sharp (1999) gave an overview of methods for allowing heterogeneity between studies. See also Higgins *et al.* (2009) for a discussion of the use of random effects, and extensions. Higgins and Thompson (2002) derived alternative measures to that in (10.4) for quantifying heterogeneity, including the I^2 statistic, the proportion of the total variation in the effect estimate due to heterogeneity between studies. See Higgins (2008) for a commentary on the quantifying of heterogeneity.

Section 10.4.3. Olkin and Sampson (1998) and Mathew and Nordstrom (1999, 2010) have commented on the equivalence of the one-step and two-step approaches to the meta-analysis of individual patient data, focusing on linear regression analyses. Stukel *et al.* (2001) compared the one-step and two-step approaches in the context of combining estimates from case-control studies when individual patient data is available. See Bowden *et al.* (2011) for similar comparisons using time-to-event data.

Section 10.6. Austin *et al.* (1997) and Martin and Austin (2000) described exact methods that can be used to combine estimates from case-control and cohort studies. See Moreno *et al.* (1996) and Huberman and Langholz (1999) for a discussion on combining matched and unmatched case-control studies.

Appendix

A theoretical diversion

A.1 Preliminaries

One of the main statistical methods used throughout the book is the method of maximum likelihood, including extensions of that method that are fruitful when only modified versions of the likelihood are available. While it would be out of place to give a detailed account of the underlying theory, we now outline the key aspects, in particular to show what is involved in the extensions.

Suppose first that we observe a single random variable Y_j with probability distribution (if discrete) or probability density function (if continuous) $f_{Y_j}(y_j; \theta)$ depending on a single parameter θ. We define the associated *score* random variable $U_j(\theta)$ by

$$U_j(\theta) = \frac{\partial \log f_{Y_j}(Y_j; \theta)}{\partial \theta}. \tag{A.1}$$

The score is a function of Y_j, and its observed value $u_j(\theta)$ is a function of the observed value y_j. Both specify how relatively rapidly the distribution changes with θ and hence convey information about θ. We define also the *information* contribution of Y_j by

$$i_j(\theta) = \text{var}\{U_j(\theta)\}. \tag{A.2}$$

Now, writing the discrete case for simplicity, we have that, summing over all possible values,

$$\sum f_{Y_j}(y_j; \theta) = 1 \tag{A.3}$$

and on differentiating with respect to θ this gives

$$\sum \frac{\partial f_{Y_j}(y_j; \theta)}{\partial \theta} = 0,$$

which on rearrangement is equivalent to

$$E\{U_j(\theta); \theta\} = \Sigma u_j(\theta) f_{Y_j}(y_j; \theta) = 0. \tag{A.4}$$

Here the notation emphasizes that the expectation is taken at the same value of θ as the derivative.

If we differentiate this last equation again with respect to θ, we have that

$$\sum u'_j(\theta) f_{Y_j}(y_j; \theta) + \sum u_j^2(\theta) f_{Y_j}(y_j; \theta) = 0,$$

which with (A.2) leads to

$$i_j(\theta) = E \left\{ -\frac{\partial^2 \log f_{Y_j}(Y_j; \theta)}{\partial \theta^2} \right\}, \qquad (A.5)$$

called the Fisher information or expected information.

Now suppose that we have n independent random variables $Y_1, \ldots, Y_n$, all with distributions depending on θ. Define the log likelihood function $l(\theta)$ as

$$l(\theta) = \sum \log f_{Y_j}(y_j; \theta), \qquad (A.6)$$

whose gradient is the total score

$$u_.(\theta) = \sum u_j(\theta) \qquad (A.7)$$

and which is such that

$$E\{U_.(\theta); \theta\} = 0. \qquad (A.8)$$

Further, the total information, the variance of the total score, is

$$i_.(\theta) = \sum i_j(\theta). \qquad (A.9)$$

Because the Y_j are assumed independent, so too are the $U_j(\theta)$ and it follows from (A.2) that

$$i_.(\theta) = \text{var}\{U_.(\theta)\}. \qquad (A.10)$$

A.2 Multidimensional parameter

Most realistic problems involve more than a single parameter. With a suitable extension of notation, the results outlined in the previous section continue to apply when the parameter θ is a vector with, say, p components $(\theta_1, \ldots, \theta_p)$. For this the score becomes a $p \times 1$ vector of partial derivatives whose rth element associated with Y_j is $u_j^r = \partial \log f_{Y_j}(y_j; \theta)/\partial \theta_r$. The information becomes a square matrix of negative second-order partial derivatives. That is, the (r, s) element of the matrix referring to Y_j is, by extension of (A.5),

$$E \left\{ -\frac{\partial^2 \log f_{Y_j}(Y_j; \theta)}{\partial \theta_r \partial \theta_s} \right\}. \qquad (A.11)$$

A direct extension of the single-parameter argument shows that for each component random variable, and for the total, because of the independence of $Y_1, \ldots, Y_n$ we have that

$$i_{rs}(\theta) = \text{cov}\{U_{.r}(\theta), U_{.s}(\theta)\}, \qquad (A.12)$$

the vector form of (A.10).

A.3 The maximum-likelihood estimate

We now return to the single-parameter case and outline the theory of the corresponding maximum-likelihood estimate, defined by

$$U_.(\hat{\theta}) = 0, \qquad (A.13)$$

an equation assumed either to have a unique solution or, if there are several stationary values of the log likelihood and hence several solutions, to have a solution that is recognizably appropriate for estimation.

We assume that there is substantial information in the data, so that the estimate $\hat{\theta}$ is close to the true value, which we now denote by θ_0. The defining equation (A.13) may then be replaced by the linear approximation

$$U_.(\theta_0) + (\hat{\theta} - \theta_0)U_.'(\theta_0) = 0. \qquad (A.14)$$

The argument now hinges on the behaviour of the terms in this equation for large n, or more strictly for large amounts of information about θ, so that errors are relatively small.

Now, $U_.(\theta_0)$ is the sum of n terms each of mean zero, and by the central limit theorem it may be assumed to be approximately normally distributed with variance $i_.(\theta_0) = n\bar{i}(\theta_0)$, say, where $\bar{i}$ is the average information per observation. That is,

$$U_.(\theta_0)/\sqrt{n}$$

has a distribution that is close to normal, with zero mean and variance not depending on n.

Next, $U'(\theta_0)$ is the sum of n terms each with non-zero mean, and therefore

$$U'(\theta_0)/n$$

is for large n close to its expectation $-\bar{i}(\theta_0)$.

We now rewrite (A.14) in the approximate form

$$\sqrt{n}(\hat{\theta} - \theta_0) = \frac{U_.(\theta_0)/\sqrt{n}}{i_.(\theta_0)/n}. \tag{A.15}$$

On the right-hand side the denominator is approximately a non-zero constant $\bar{i}(\theta_0)$ and the numerator is approximately a normally distributed random variable of zero mean and variance $\bar{i}(\theta_0)$.

It can be shown from this that $\sqrt{n}(\hat{\theta} - \theta_0)$ is approximately normally distributed with variance $\bar{i}(\theta_0)/\bar{i}^2(\theta_0)$, that is, on restoring the dependence on n the maximum likelihood estimate $\hat{\theta}$ is approximately normally distributed whose mean is the true value θ_0 and whose variance is $i^{-1}(\theta_0)$.

In this argument the dependence on n has been shown explicitly in order to indicate how the final conclusion is reached; in applications either we use the total expected information $i_.$, typically evaluated at $\hat{\theta}$, or, more commonly, we replace the expected information by its observed counterpart evaluated at $\hat{\theta}$,

$$j(\hat{\theta}) = -\frac{\partial^2 l(\hat{\theta})}{\partial \theta^2}. \tag{A.16}$$

Note particularly that in this derivation the information $i_.(\theta_0)$ enters in two distinct ways, in the variance of the score and also in the curvature of the log likelihood; a cancellation occurs to produce the final answer.

When the parameter is multidimensional then the approximation (A.14) to the defining equation (A.13) becomes a set of simultaneous linear equations and the solution can be written in the form, omitting the factors determined by n,

$$\hat{\theta} - \theta_0 = i_.^{-1} U_.(\theta_0), \tag{A.17}$$

where $i_.^{-1}$ is the inverse of the information matrix. It follows that the maximum-likelihood estimate is approximately normally distributed with mean θ_0 and, in the vector-parameter case, the covariance matrix

$$i_.^{-1} i_. i_.^{-1} = i_.^{-1}. \tag{A.18}$$

In particular, the standard error of the estimate of a particular component can now be found, using the appropriate diagonal element of (A.18) or its observed counterpart j, and confidence limits calculated.

A.4 Significance tests

Throughout the book we have emphasized the calculation of estimates and
their standard errors for the effects of interest, often log odds ratios or
similar quantities. Equivalently these give approximate confidence intervals. The point estimates and standard errors are typically calculated by
the methods of Section A.3. For testing null hypotheses there are closely
related methods based directly on the log likelihood function or sometimes
on the score. These have the advantage, in particular, of being unchanged
by transformations of the parameter, for example from a log odds ratio to
an odds ratio. We do not discuss these alternative approaches in the present
book.

A.5 More detailed theory

Although the essence of the mathematical discussion of the properties of
maximum-likelihood estimates is given above, much more detail is necessary for a mathematically complete account of the behaviour of maximum-
likelihood estimates as n becomes large. To study the behaviour of the
estimates in samples of typical size requires a mixture of further theory,
comparison with situations in which the exact theory is available and simulation. Provided that the numbers of cases in different groups are not small
it is likely that most of the large-sample methods described in this book
will be adequate; the presence of, for example, an appreciable proportion
of zero-outcome frequencies would certainly call for some modification.

A.6 Pseudo-likelihood

The results sketched above derive from the log likelihood function of the
data, that is, the log probability of what is observed, as specified directly
by the probability model chosen as an idealized model representing the
data-generating process. The likelihood is typically a product of factors.
For independent individuals there is one from each individual under study,
and the likelihood is considered as a function of the unknown parameters.
In some applications considered in this book, especially where the times of
occurrence of cases are involved, similar procedures are used in which the
log likelihood is replaced by a function possibly less directly connected
with the data-generating process. We call such log likelihoods *pseudo
log likelihoods*; other terms sometimes used include 'derived' or 'composite' likelihoods. Typically each factor is a log probability but is often

conditioned on some aspect of the stochastic history of the individual. Do the above results apply and if not, how are they to be modified?

There are essentially two aspects to consider. First, is the distribution of the estimate correctly centred and hence not subject to major systematic error? For this it is enough that the score for each individual, defined as that before as a log likelihood derivative, has expectation zero. If each factor is a proper probability then the previous argument, and so the required property, holds.

The second requirement is that the variance of the total score can be determined fairly directly from log likelihood derivatives. Typically the previous argument applies to each factor of the likelihood, but there is the possibility that the different components of the score function are correlated, for example because of the time dependences involved. The key general result is that in the single-parameter case

$$\text{var}\{U_.(\theta)\} = \sum \text{var}\{U_j(\theta)\} + 2\sum_{k>j} \text{cov}\{U_j(\theta), U_k(\theta)\}, \quad \text{(A.19)}$$

and the final result is that the asymptotic covariance matrix of $\hat{\theta}$ is

$$i^{-1}\text{cov}(U_.)i^{-1}, \quad \text{(A.20)}$$

often called the *sandwich formula*.

If and only if the covariance contributions in the middle factor are zero or can be ignored then the simpler version i^{-1} is recovered.

Now, in the extended version of the likelihood considered in Chapters 7 and 8 each contribution, U_k, say, to the score is conditioned on a set of events, a history $\mathcal{H}_k$, say. If the contributions to the total score can be ordered as $1, \ldots, n$ in such a way that $\mathcal{H}_k$ contains the values of the score at previous points j, $j < k$, then, provided the contributions are unbiased, we have that

$$E(U_k | U_j = u_j, j < k) = 0. \quad \text{(A.21)}$$

We multiply by the constant value of u_j and then take the expectation over its value to obtain

$$E(U_j U_k) = \text{cov}(U_j, U_k) = 0. \quad \text{(A.22)}$$

This, which is in effect the nesting of the successive conditioning histories $\mathcal{H}_j$, is enough to ensure the applicability of the simpler formula for the asymptotic covariance matrix. Otherwise the score covariances should be calculated and the sandwich formula invoked.

When the covariance matrix of the U_j departs appreciably from the simple form, modification of the estimating equation $U.(\hat{\theta}) = 0$ may be desirable.

A.7 Weighted estimation

Some more complicated forms of design in effect involve sampling in which different individuals have different and known probabilities of selection. To estimate a sum over the corresponding population, the bias induced by the differential probabilities of selection is corrected by the Horvitz–Thompson formula. The essence of the theory is as follows. Suppose that a notional population of individuals has values $\{y_1, \ldots, y_N\}$ and that a sample of n individuals $Y_1, \ldots, Y_n$ is chosen from the population, each individual being selected according to a set of probabilities, namely $\{\pi_1, \ldots, \pi_N\}$. We now form derived observations in which each observed Y_i is weighted inversely by its probability of selection, that is, we set $Y_i^* = Y_i/\pi_s$ whenever $Y_i = y_s$. Then $E(Y_i^*) = \Sigma y_s$, so that an unbiased estimate of the finite population total is obtained.

A.8 Bayesian analysis

The discussion of statistical inference in this book has been framed largely in terms of procedures based on the method of maximum likelihood or an adaptation thereof, with occasional reference to small-sample exact procedures. In all these procedures the parameters describing the data-generating process are regarded as unknown constants. An alternative approach, broadly described as Bayesian, assigns the unknown parameters a probability distribution, called the *prior distribution*. This encapsulates what is known about the unknown parameter based on considerations excluding the data under analysis. The prior distribution is then combined with the likelihood function derived from the model and data to produce the *posterior distribution* of the parameters, amalgamating both sources of information. We shall not discuss Bayesian approaches to case-control studies in depth but some general points need to be made.

First, the nature and purpose of the prior distribution are critical. There are a number of distinct possibilities.

(1) The prior distribution may encapsulate evidence-based external information, either from background knowledge or from the summarization of previous somewhat similar studies. In the latter case it would be in

principle desirable to analyse all the available data as a whole, but this may not be feasible. In the former case the prior distribution may be interpreted as an equivalent to the information in a particular configuration of data, for example a notional study of a specified size and form.

(2) The prior distribution may be very dispersed, the intention being to express the absence of external information, so that the posterior distribution then summarizes what can be learned from the data. The numerical answers from this approach can often be made to agree exactly or to a close approximation with the non-Bayesian approach, when the number of parameters is not too large, but the use of such so-called flat or uninformative priors is dangerous if the dimension of the parameter space is large.

(3) A third, and uncontroversial, approach arises if the study in question is one of a considerable number of broadly similar studies in which the parameter of interest varies randomly from study to study. This leads to a so-called random or hierarchical effects model. It becomes a Bayesian analysis in the stricter sense if interest is focused on the parameter in a single study or possibly on a contrast of the values of the parameter in different studies. A further development of this second possibility arises if interest then returns to the parameter of interest in an individual study. There are now two sources of information, the particular study in question and the collective of studies of which it is a member. The use of information in the latter is sometimes referred to as 'borrowing strength'. Provided that the distribution across studies is not very long-tailed, a formal application of the Bayesian approach justifies moving the estimate from the individual study towards the general mean.

We now consider the first two possibilities in turn.

A.8.1 Informative prior distribution

The population model underlying the simplest case-control study is characterized by a distribution of the exposure variables in the population and by parameters specifying the probability of an individual's being a case as a function of the exposure variables. A complete Bayesian analysis requires the specification of a prior distribution of the parameters. This is multiplied by the likelihood stemming from the case-control data, leading to the posterior distribution of the parameters of interest, namely those expressing the dependence of the probability of being a case on the exposure measures.

Seaman and Richardson (2004) developed this argument in the special case of discrete exposure measures.

Here we bypass most of the details by the following argument. Suppose that the parameter of interest is ψ and the nuisance parameters are collectively given by λ. Then in large samples the likelihood is proportional to a Gaussian density centred on $(\hat{\psi}, \hat{\lambda})$ with inverse covariance matrix, say j, which is the matrix of minus the second derivatives of the log likelihood at the maximum. If now the prior density is Gaussian with mean (m_ψ, m_λ) and inverse covariance matrix j_0, it follows, on multiplying the prior by the likelihood and renormalizing, that the posterior density of (ψ, λ) is asymptotically multivariate Gaussian, with mean

$$\{(\hat{\psi}, \hat{\lambda})j + (m_\psi, m_\lambda)j_0\}(j + j_0)^{-1} \qquad (A.23)$$

and covariance matrix $(j + j_0)^{-1}$.

Although it is not part of the formal procedure to specify this aspect, the merging of information from the data and the prior is sensible only when the two sources of information are reasonably consistent, in particular when m_ψ and $\hat{\psi}$ are not too different.

Further, especially if the matrices j, j_0 are far from diagonal, a conclusion about ψ may be affected by information about λ contained in the prior. In other cases, however, it may be helpful to simplify the problem by an approximation that uses only $\hat{\psi}$, the relevant information measure $j_{\psi\psi.\lambda}$ and the prior for ψ with mean m_ψ and information $j_{0,\psi}$. Then the posterior distribution for ψ, calculated as if that were the only information available, has as its mean the weighted mean of $\hat{\psi}$ and m_ψ and as its variance $(j_{\psi\psi.\lambda} + j_{0,\psi})^{-1}$.

One interpretation of the amount of information supplied by the prior is in terms of the equivalent number of independent observations.

A.8.2 Uninformative priors

In the previous discussion if the amount of data for analysis is large and the prior information limited then the data-based contribution to the posterior distribution of ψ will be dominant. If the intention is solely to isolate the contribution of the data then it may be asked whether a prior is available that would contain no information about the parameter of concern. Note that some notion of a prior is needed if we are to obtain a *probability* distribution for the unknown parameter, which in the main approach we have adopted is treated simply as an unknown constant. There has been a long dispute over whether such a notion of a perfectly uninformative prior is viable.

Clearly it must be highly dispersed, but is that the same as saying there is no information?

An alternative view is to consider what prior would produce as closely as possible the numerical results of confidence-interval theory. For a single parameter this is possible by in effect transforming the parameter to be in a form close to a location parameter and then assigning a uniform prior over a very wide interval. Similar special arguments do not in general hold when there is more than one unknown parameter; more importantly, if there are a large number of parameters then the use of flat priors may lead to dangerously misleading conclusions.

For a case-control study with exposure variables having discrete values, Seaman and Richardson (2004) showed that numerical results equivalent to those given in Chapter 4 are obtained as limiting cases when ψ has a uniform distribution over a wide interval and the prior distribution in the population over the exposure characteristics is very irregular.

Notes

See Cox (2006) for a general discussion of the issues in the appendix and for further references to the very extensive literature on the theoretical aspects of statistical inference. Pseudo-likelihoods were introduced by Besag (1974, 1977). For the sandwich formula, see Godambe (1960) and Cox (1961, 1962). For the Horvitz–Thompson estimate, see Thompson (1992).

References

Aalen, O.O. 1989. A linear regression model for the analysis of life times. *Statistics in Medicine*, **8**, 907–925. (Cited on p. 189.)

Agresti, A. 1984. *Analysis of Ordinal Categorical Data*. New York: Wiley. (Cited on p. 131.)

Agresti, A. 1990. *Categorical Data Analysis*. New York: Wiley. (Cited on p. 131.)

Andersen, P.K., Borgan, Ø., Gill, R., and Keiding, N. 1993. *Statistical Models Based on Counting Processes*. New York: Springer Verlag. (Cited on p. 187.)

Anderson, J.A. 1972. Separate sample logistic discrimination. *Biometrika*, **59**, 19–35. (Cited on pp. 30 and 108.)

Andrieu, N., Goldstein, A.M., Thomas, D.C., and Langholz, B. 2001. Counter-matching in studies of gene–environment interaction: efficiency and feasibility. *American Journal of Epidemiology*, **153**, 265–274. (Cited on p. 188.)

Anzures-Cabrera, J. and Higgins, J.P. 2010. Graphical displays for univariate meta-analysis: an overview with suggestions for practice. *Research Synthesis Methods*, **1**, 66–80. (Cited on p. 251.)

Aranda-Ordaz, F.J. 1983. An extension of the proportional hazards model for grouped data. *Biometrics*, **39**, 109–117. (Cited on p. 188.)

Armitage, P. 1975. The use of the cross-ratio in aetiological surveys, in *Perspectives in Probability and Statistics*, pp. 349–355. London: Academic Press. (Cited on p. 109.)

Armstrong, B., Tremblay, C., Baris, D., and Theriault, G. 1994. Lung cancer mortality and polynuclear aromatic hydrocarbons: a case-cohort study of aluminum production workers in Arvida, Quebec, Canada. *American Journal of Epidemiology*, **139**, 250–262. (Cited on p. 194.)

Armstrong, B.G., Whittemore, A.S., and Howe, G.R. 1989. Analysis of case-control data with covariate measurement error: application to diet and colon cancer. *Statistics in Medicine*, **8**, 1151–1163. (Cited on p. 238.)

Austin, H. and Flanders, W.D. 2003. Case-control studies of genotypic relative risks using children of cases as controls. *Statistics in Medicine*, **22**, 129–145. (Cited on p. 158.)

Austin, H., Perkins, L.L., and Martin, D.O. 1997. Estimating a relative risk across sparse case-control and follow-up studies: a method for meta-analysis. *Statistics in Medicine*, **16**, 1005–1015. (Cited on p. 251.)

Barlow, R.E., Bartholomew, D.J., Bremner, J.M., and Brunk, H.D. 1972. *Statistical Inference Under Order Restrictions*. New York: Wiley. (Cited on p. 131.)

262

Barlow, W.E. 1994. Robust variance estimation for the case-cohort design. *Biometrics*, **50**, 1064–1072. (Cited on p. 210.)

Barlow, W.E. and Prentice, R.L. 1988. Residuals for relative risk regression. *Biometrika*, **75**, 65–74. (Cited on p. 210.)

Barlow, W.E., Ichikawa, L., Rosner, D., and Izumi, S. 1999. Analysis of case-cohort designs. *Journal of Clinical Epidemiology*, **52**, 1165–1172. (Cited on p. 210.)

Barnard, G.A. 1949. Statistical inference (with discussion). *Journal of the Royal Statistical Society B*, **11**, 115–149. (Cited on p. 59.)

Barron, B.A. 1977. The effects of misclassification on the estimation of relative risk. *Biometrics*, **33**, 414–418. (Cited on p. 236.)

Becher, H. 1991. Alternative parametrization of polychotomous models: theory and application to matched case-control studies. *Statistics in Medicine*, **10**, 375–382. (Cited on p. 131.)

Begg, C.B. and Gray, R. 1984. Calculation of polychotomous logistic regression parameters using individualized regressions. *Biometrika*, **71**, 11–18. (Cited on p. 130.)

Benichou, J. 1991. Methods of adjustment for estimating the attributable risk in case-control studies: a review. *Statistics in Medicine*, **10**, 1753–1773. (Cited on p. 110.)

Benichou, J. and Gail, M.H. 1990. Variance calculations and confidence intervals for estimates of the attributable risk based on logistic models. *Biometrics*, **46**, 991–1003. (Cited on p. 109.)

Benichou, J. and Gail, M.H. 1995. Methods of inference for estimates of absolute risk derived from population-based case-control studies. *Biometrics*, **51**, 182–194. (Cited on p. 60.)

Benichou, J., Byrne, C., and Gail, M.H. 1997. An approach to estimating exposure-specific rates of breast cancer from a two-stage case-control study within a cohort. *Statistics in Medicine*, **16**, 133–151. (Cited on p. 150.)

Berkson, J. 1950. Are there two regressions? *Journal of the American Statistical Association*, **45**, 164–180. (Cited on p. 237.)

Bernstein, J.L., Langholz, B., Haile, R.W., Bernstein, L., Thomas, D.C., Stovall, M. *et al.* 2004. Study design: evaluating gene–environment interacions in the etiology of breast cancer – the WECARE study. *Breast Cancer Research*, **6**, R199–R214. (Cited on p. 173.)

Berzuini, C., Dawid, A.P., and Bernardinelli, L. (editors) 2012. *Causality*. Chichester: Wiley. (Cited on p. 61.)

Besag, J.E. 1974. Spatial interaction and the statistical analysis of lattice systems (with discussion). *Journal of the Royal Statistical Society B*, **36**, 192–236. (Cited on p. 261.)

Besag, J.E. 1977. Efficiency of pseudo-likelihood estimates for simple Gaussian fields. *Biometrika*, **64**, 616–618. (Cited on p. 261.)

Bishop, Y.M., Fienberg, S.E., and Holland, P.W. 1975. *Discrete Multivariate Analysis: Theory and Practice*. Cambridge, Mass.: MIT Press. (Cited on p. 60.)

Borenstein, M., Hedges, L.V., Higgins, J.P.T., and Rothstein, H.R. 2010. A basic introduction to fixed-effect and random-effects models for meta-analysis. *Research Synthesis Methods*, **1**, 97–111. (Cited on p. 251.)

Borgan, Ø. and Langholz, B. 1993. Nonparametric estimation of relative mortality from nested case-control studies. *Biometrics*, **49**, 593–602. (Cited on p. 189.)

Borgan, Ø. and Langholz, B. 1997. Estimation of excess risk from case-control data using Aalen's linear regression model. *Biometrics*, **53**, 690–697. (Cited on p. 189.)

Borgan, Ø. and Olsen, E.F. 1999. The efficiency of simple and counter-matched nested case-control sampling. *Scandinavian Journal of Statistics*, **26**, 493–509. (Cited on pp. 188 and 189.)

Borgan, Ø., Goldstein, L., and Langholz, B. 1995. Methods for the analysis of sampled cohort data in the Cox proportional hazards model. *Annals of Statistics*, **23**, 1749–1778. (Cited on pp. 187, 188 and 189.)

Borgan, Ø., Langholz, B., Samuelsen, S.O., Goldstein, L., and Pogoda, J. 2000. Exposure stratified case-cohort designs. *Lifetime Data Analysis*, **6**, 39–58. (Cited on p. 210.)

Bowden, J., Tierney, J.F., Simmonds, M., Copas, A.J., and Higgins, J.P. 2011. Individual patient data meta-analysis of time-to-event outcomes: one-stage versus two-stage approaches for estimating the hazard ratio under a random effects model. *Research Synthesis Methods*, **2**, 150–162. (Cited on p. 251.)

Bradford Hill, A. 1965. The environment and disease: association or causation? *Proceedings of the Royal Society of Medicine*, **58**, 295–300. (Cited on p. 61.)

Breslow, N.E. 1972. Contribution to the discussion of the paper by D.R. Cox. *Journal of the Royal Statistical Society B*, **34**, 216–217. (Cited on p. 189.)

Breslow, N.E. 1981. Odds ratio estimators when the data are sparse. *Biometrika*, **68**, 73–84. (Cited on p. 61.)

Breslow, N.E. 1996. Statistics in epidemiology: the case-control study. *Journal of the American Statistical Association*, **91**, 14–28. (Cited on p. 29.)

Breslow, N.E. and Cain, K.C. 1988. Logistic regression for two-stage case-control data. *Biometrika*, **75**, 11–20. (Cited on pp. 157 and 158.)

Breslow, N.E. and Chatterjee, N. 1999. Design and analysis of two-phase studies with binary outcome applied to Wilms tumour prognosis. *Applied Statistics*, **48**, 457–468. (Cited on pp. 157 and 158.)

Breslow, N.E. and Day, N.E. 1980. *Statistical Methods in Cancer Research: Volume 1 – The Analysis of Case-Control Studies*. Lyons: International Agency for Research on Cancer. (Cited on pp. 29 and 61.)

Breslow, N.E. and Holubkov, R. 1997. Maximum likelihood estimation of logistic regression parameters under two-phase, outcome-dependent sampling. *Journal of the Royal Statistical Society B*, **59**, 447–461. (Cited on pp. 157 and 158.)

Breslow, N.E. and Liang, K.Y. 1982. The variance of the Mantel–Haenszel estimator. *Biometrics*, **38**, 943–952. (Cited on p. 61.)

Breslow, N.E. and Powers, W. 1978. Are there two logistic regressions for retrospective studies? *Biometrics*, **34**, 100–105. (Cited on p. 108.)

Breslow, N.E. and Zhao, L.P. 1988. Logistic regression for stratified case-control studies. *Biometrics*, **44**, 891–899. (Cited on p. 157.)

Breslow, N.E., Day, N.E., Halvorsen, K.T., Prentice, R.L., and Sabai, C. 1978. Estimation of multiple relative risk functions in matched case-control studies. *American Journal of Epidemiology*, **108**, 299–307. (Cited on p. 109.)

Breslow, N.E., Lumley, T., Ballantyne, C.M., Chambless, L.E., and Kulich, M. 2009a. Improved Horvitz–Thompson estimation of model parameters from two-phase stratified samples: applications in epidemiology. *Statistics in Biosciences*, **1**, 32–49. (Cited on p. 211.)

Breslow, N.E., Lumley, T., Ballantyne, C.M., Chambless, L.E., and Kulich, M. 2009b. Using the whole cohort in the analysis of case-cohort data. *American Journal of Epidemiology*, **169**, 1398–1405. (Cited on p. 211.)

Bruzzi, P., Green, S.B., Byar, D.P., Brinton, L.A., and Shairer, C. 1985. Estimating the population attributable risk for multiple risk factors using case-control data. *American Journal of Epidemiology*, **122**, 904–914. (Cited on p. 110.)

Bull, S.B. and Donner, A. 1993. A characterization of the efficiency of individualized logistic regressions. *The Canadian Journal of Statistics*, **21**, 71–78. (Cited on p. 130.)

Buonaccorsi, J.P. 2010. *Measurement Error. Models, Methods and Applications*. Chapman & Hall/CRC. (Cited on p. 236.)

Cain, K.C. and Breslow, N.E. 1988. Logistic regression analysis and efficient design for two-stage studies. *American Journal of Epidemiology*, **128**, 1198–1206. (Cited on p. 157.)

Carroll, R.J., Spiegelman, C., Lan, K.K, Bailey, K.T., and Abbott, R.D. 1984. On errors-in-variables for binary regression models. *Biometrika*, **71**, 19–26. (Cited on p. 238.)

Carroll, R.J., Gail, M.H., and Lubin, J.H. 1993. Case-control studies with errors in covariates. *Journal of the American Statistical Association*, **88**, 185–199. (Cited on pp. 237 and 239.)

Carroll, R.J., Wang, S., and Wang, C.Y. 1995. Prospective analysis of logistic case-control studies. *Journal of the American Statistical Association*, **90**, 157–169. (Cited on p. 238.)

Carroll, R.J., Ruppert, D., Stefanski, L.A., and Crainiceanu, C.M. 2006. *Measurement Error in Nonlinear Models: A Modern Perspective*. 2nd edn. Chapman & Hall/CRC. (Cited on pp. 236, 237 and 238.)

Chatterjee, N. and Carroll, R.J. 2005. Semiparametric maximum likelihood estimation exploiting gene-environment independence in case-control studies. *Biometrika*, **92**, 399–418. (Cited on pp. 158 and 159.)

Chen, H.Y. and Little, R.J.A. 1999. Proportional hazards regression with missing co-variates. *Journal of the American Statistical Association*, **94**, 896–908. (Cited on p. 190.)

Chen, T. 2001. RE: Methods of adjustment for estimating the attributable risk in case-control studies; a review. *Statistics in Medicine*, **20**, 979–982. (Cited on p. 110.)

Chen, Y.-H. 2002. Cox regression in cohort studies with validation sampling. *Journal of the Royal Statistical Society B*, **64**, 51–62. (Cited on p. 211.)

Chen, Y.T., Dubrow, R., Zheng, T., Barnhill, R.L., Fine, J., and Berwick, M. 1998. Sunlamp use and the risk of cutaneous malignant melanoma: a population-based case-control study in Connecticut, USA. *International Journal of Epidemiology*, **27**, 758–765. (Cited on pp. 27 and 41.)

Cheng, C.L. and Schneeweiss, H. 1998. Polynomial regression with errors in the variables. *Journal of the Royal Statistical Society B*, **60**, 189–199. (Cited on p. 238.)

Chu, R., Gustafson, P., and Le, N. 2010. Bayesian adjustment for exposure misclassification in case-control studies. *Statistics in Medicine*, **29**, 994–1003. (Cited on p. 236.)

Cochran, W.G. 1954. The combination of estimates from different experiments. *Biometrics*, **10**, 101–129. (Cited on p. 251.)

Cochran, W.G. 1965. The planning of observational studies of human populations (with discussion). *Journal of the Royal Statistical Society A*, **128**, 234–265. (Cited on pp. 61 and 81.)

Cole, S.R., Chu, H., and Greenland, S. 2006. Multiple-imputation for measurement error correction. *International Journal of Epidemiology*, **35**, 1074–1081. (Cited on p. 238.)

Cologne, J.B. and Langholz, B. 2003. Selecting controls for assessing interaction in nested case-control studies. *Journal of Epidemiology*, **13**, 193–202. (Cited on p. 188.)

Cologne, J.B., Sharp, G.B., Neriishi, K., Verkasalo, P.K., Land, C.E., and Nakachi, K. 2004. Improving the efficiency of nested case-control studies of interaction by selection of controls using counter-matching on exposure. *International Journal of Epidemiology*, **33**, 485–492. (Cited on p. 188.)

Cook, J.R. and Stefanski, L.A. 1994. Simulation-extrapolation estimation in parametric measurement error models. *Journal of the American Statistical Association*, **89**, 1314–1328. (Cited on p. 237.)

Cornfield, J. 1951. A method of estimating comparative rates from clinical data. Applications to cancer of the lung, breast and cervix. *Journal of the National Cancer Institute*, **11**, 1269–1275. (Cited on pp. 29 and 30.)

Cornfield, J., Haenszel, W., Hammond, E.C., Lilienfeld, A.M., Shimkin, M.B., and Wynder, E.L. 1959. Smoking and lung cancer: recent evidence and a discussion of some questions. *Journal of the National Cancer Institute*, **22**, 173–203. Reprinted in *International Journal of Epidemiology* **38** (2009), 1175–1191. (Cited on pp. 61 and 251.)

Cosslett, S. 1981. Maximum likelihood estimators for choice-based samples. *Econometrica*, **49**, 1289–1316. (Cited on pp. 108 and 157.)

Cox, D.R. 1955. Some statistical methods connected with series of events (with discussion). *Journal of the Royal Statistical Society B*, **17**, 129–164. (Cited on p. 60.)

Cox, D.R. 1958a. The regression analysis of binary sequences (with discussion). *Journal of the Royal Statistical Society B*, **20**, 215–242. (Cited on p. 61.)

Cox, D.R. 1958b. Two further applications of a model for binary regression. *Biometrika*, **45**, 562–565. (Cited on p. 82.)

Cox, D.R. 1961. Tests of separate families of hypotheses, in *Proc. 4th Berkeley Symp.*, **1**, pp. 105–123. (Cited on p. 261.)

Cox, D.R. 1962. Further results on tests of separate families of hypotheses. *Journal of the Royal Statistical Society B*, **24**, 406–424. (Cited on p. 261.)

Cox, D.R. 1966. Some procedures connected with the logistic qualitative response curve, in Festschrift for J. Neyman, pp. 55–71. London: Wiley. (Cited on p. 238.)

Cox, D.R. 1972. Regression models and life tables. *Journal of the Royal Statistical Society Series B*, **34**, 187–202. (Cited on pp. 186 and 189.)

Cox, D.R. 1975. Partial likelihood. *Biometrika*, **62**, 269–276. (Cited on p. 186.)

Cox, D.R. 2006. *Principles of Statistical Inference*. Cambridge: Cambridge University Press. (Cited on p. 261.)

Cox, D.R. and Lewis, P.A.W. 1966. *The Statistical Analysis of Series of Events*. London: Methuen. (Cited on p. 131.)

Cox, D.R. and Oakes, D. 1984. *Analysis of Survival Data*. London: Chapman and Hall. (Cited on p. 187.)

Cox, D.R. and Snell, E.J. 1989. *Analysis of Binary Data*. 2nd edn. London: Chapman and Hall. (Cited on pp. 61, 82 and 109.)

Cox, D.R. and Wermuth, N. 2004. Causality: a statistical view. *International Statistical Review*, **72**, 285–305. (Cited on p. 61.)

Cox, D.R. and Wong, M.Y. 2004. A simple procedure for the selection of significant effects. *Journal of the Royal Statistical Society B*, **66**, 395–400. (Cited on p. 188.)

Dahm, C.C., Keogh, R.H., Spencer, E.A., Greenwood, D.C., Key, T.J., Fentiman, I.S. *et al.* 2010. Dietary fiber and colorectal cancer risk: a nested casecontrol study using food diaries. *Journal of the National Cancer Institute*, **102**, 614–626. (Cited on p. 65.)

Darby, S.C., Ewertz, M., McGale, P., Bennet, A.M., Blom-Goldman, U., Brønnum, D. *et al.* 2013. Risk of ischaemic heart disease in women after radiotherapy for breast cancer. *New England Journal of Medicine*, **368**, 987–998. (Cited on p. 177.)

Day, N.E., Oakes, S., Luben, R., Khaw, K.T., Bingham, S., Welch, A. *et al.* 1999. EPIC in Norfolk: study design and characteristics of the cohort. *British Journal of Cancer*, **80** (Suppl. 1), 95–103. (Cited on p. 185.)

De Stavola, B.L. and Cox, D.R. 2008. On the consequences of overstratification. *Biometrika*, **95**, 992–996. (Cited on p. 109.)

DerSimonian, R. and Laird, N. 1986. Meta-analysis in clinical trials. *Controlled Clinical Trials*, **7**, 177–188. (Cited on p. 251.)

Didelez, V., Kreiner, S., and Keiding, N. 2010. On the use of graphical models for inference under outcome-dependent sampling. *Statistical Science*, **25**, 368–387. (Cited on pp. 7, 30 and 61.)

Doll, R. and Bradford Hill, A. 1950. Smoking and carcinoma of the lung. *British Medical Journal*, **2**, 739–748. (Cited on p. 26.)

Domowitz, I. and Sartain, R.L. 1999. Determinants of the consumer bankruptcy decision. *Journal of Finance*, **54**, 403–420. (Cited on p. 28.)

Drescher, K. and Schill, W. 1991. Attributable risk estimation from case-control data via logistic regression. *Biometrics*, **47**, 1247–1256. (Cited on p. 110.)

Dubin, N. and Pasternack, B.S. 1986. Risk assessment for case-control subgroups by polychotomous logistic regression. *American Journal of Epidemiology*, **123**, 1101–1117. (Cited on p. 130.)

Duffy, S.W., Rohan, T.E., Kandel, R., Prevost, T.C., Rice, K., and Myles, J.P. 2003. Misclassification in a matched case-control study with variable matching ratio: application to a study of c-erbB-2 overexpression and breast cancer. *Statistics in Medicine*, **22**, 2459–2468. (Cited on p. 237.)

Duffy, S.W, Warwick, J., Williams, A.R.W., Keshavarz, H., Kaffashian, F., Rohan, T.E. *et al.* 2004. A simple model for potential use with a misclassified binary outcome in epidemiology. *Journal of Epidemiology and Community Health*, **58**, 712–717. (Cited on p. 236.)

Dupont, W.D. 1988. Power calculations for matched case-control studies. *Biometrics*, **44**, 1157–1168. (Cited on p. 82.)

Easton, D.J., Peto, J., and Babiker, A.G. 1991. Floating absolute risk: alternative to relative risk in survival and case-control analysis avoiding an arbitrary reference group. *Statistics in Medicine*, **10**, 1025–1035. (Cited on p. 61.)

Edwards, A.W.F. 1963. The measure of information in a 2×2 table. *Journal of the Royal Statistical Society A*, **126**, 109–114. (Cited on pp. 30 and 59.)

Farewell, V.T. 1979. Some results on the estimation of logistic models based on retrospective data. *Biometrika*, **66**, 27–32. (Cited on pp. 30 and 108.)

Farrington, C.P. 1995. Relative incidence estimation from case series for vaccine safety evaluation. *Biometrics*, **51**, 228–235. (Cited on p. 131.)

Fearn, T., Hill, D.C., and Darby, S.C. 2008. Measurement error in the explanatory variable of a binary regression: regression calibration and integrated conditional likelihood in studies of residential random and lung cancer. *Statistics in Medicine*, **27**, 2159–2176. (Cited on pp. 97 and 239.)

Fears, T.R. and Brown, C.C. 1986. Logistic regression methods for retrospective case-control studies using complex sampling procedures. *Biometrics*, **42**, 955–960. (Cited on p. 157.)

Feinstein, A.R. 1987. Quantitative ambiguities on matched versus unmatched analyses of the 2×2 table for a case-control study. *International Journal of Epidemiology*, **16**, 128–134. (Cited on p. 109.)

Firth, D. and de Menezes, R.X. 2004. Quasi-variances. *Biometrika*, **91**, 65–80. (Cited on p. 61.)

Fisher, R.A. 1922. On the interpretation of chi square from contingency tables and the calculation of P^*. *Journal of the Royal Statistical Society*, **85**, 87–94. (Cited on p. 60.)

Fisher, R.A. 1935. The logic of inductive inference (with discussion). *Journal of the Royal Statistical Society*, **98**, 39–82. (Cited on p. 60.)

Flanders, W.D. and Greenland, S. 1991. Analytic methods for two-stage case-control studies and other stratified designs. *Statistics in Medicine*, **10**, 739–747. (Cited on p. 158.)

Flanders, W.D., Dersimonian, R., and Rhodes, P. 1990. Estimation of risk ratios in case-base studies with competing risks. *Statistics in Medicine*, **9**, 423–435. (Cited on p. 209.)

Freedman, L.S., Fainberg, V., Kipnis, V., Midthune, D., and Carroll, R.J. 2004. A new method for dealing with measurement error in explanatory variables of regression models. *Biometrics*, **60**, 172–181. (Cited on p. 238.)

Freedman, L.S., Midthune, D., Carroll, R.J., and Kipnis, V. 2008. A comparison of regression calibration, moment reconstruction and imputation for adjusting for covariate measurement error in regression. *Statistics in Medicine*, **27**, 5195–5216. (Cited on p. 238.)

Gail, M.H. 1998. Case-control study, hospital-based, in *Encyclopedia of Biostatistics, Volume 1*, p. 514. Chichester: Wiley. (Cited on p. 30.)

Gail, M.H., Wieand, S., and Piantadosi, S. 1984. Biased estimates of treatment effect in randomized experiments with nonlinear regression and omitted covariates. *Biometrika*, **71**, 431–444. (Cited on pp. 52, 61 and 109.)

Gatto, N.M., Campbell, U.B., Rundle, A.G., and Ahsan, H. 2004. Further development of the case-only design for assessing gene–environment interaction: evaluation of and adjustment for bias. *International Journal of Epidemiology*, **33**, 1014–1024. (Cited on p. 131.)

Gauderman, W.J., Witte, J.S., and Thomas, D.C. 1999. Family-based association studies. *Journal of the National Cancer Institute Monographs*, **26**, pp. 31–37. (Cited on p. 158.)

Gebregziabher, M., Guimaraes, P., Cozen, W., and Conti, D.V. 2010. A polytomous conditional likelihood approach for combining matched and unmatched case-control studies. *Statistics in Medicine*, **29**, 1004–1013. (Cited on p. 131.)

Godambe, V.P. 1960. An optimum property of regular maximum likelihood estimation. *Annals of Mathematical Statistics*, **31**, 1208–1212. (Cited on p. 261.)

Goldstein, L. and Langholz, B. 1992. Asymptotic theory for nested case-control sampling in the Cox regression model. *Annals of Statistics*, **20**, 1903–1928. (Cited on p. 187.)

Goodman, L.A. and Kruskal, W.H. 1954. Measures of association for cross-classifications. *Journal of the American Statistical Association*, **49**, 732–764. (Cited on p. 59.)

Goodman, L.A. and Kruskal, W.H. 1959. Measures of association for cross-classifications. II. Further discussion and references. *Journal of the American Statistical Association*, **54**, 123–163. (Cited on p. 59.)

Goodman, L.A. and Kruskal, W.H. 1963. Measures of association for cross-classifications. III. Approximate sampling theory. *Journal of the American Statistical Association*, **58**, 310–364. (Cited on p. 59.)

Green, J., Berrington de Gonzalez, A., Sweetland, S., Beral, V., Chilvers, C., Crossley, B. *et al.* 2003. Risk factors for adenocarcinoma and squamous cell carcinoma of the cervix in women aged 20–44 years: the UK National Case-Control Study of Cervical Cancer. *British Journal of Cancer*, **89**, 2078–2086. (Cited on p. 95.)

Greenland, S. 1982. The effect of misclassification in matched-pair case-control studies. *American Journal of Epidemiology*, **116**, 402–406. (Cited on p. 237.)

Greenland, S. 1986. Adjustment of risk ratios in case-base studies (hybrid epidemiologic designs). *Statistics in Medicine*, **5**, 579–584. (Cited on p. 209.)

Greenland, S. 1989. On correcting for misclassification in twin studies and other matched-pair studies. *Statistics in Medicine*, **8**, 825–829. (Cited on p. 237.)

Greenland, S. and Kleinbaum, D.G. 1983. Correcting for misclassification in two-way tables and matched-pair studies. *International Journal of Epidemiology*, **12**, 93–97. (Cited on pp. 236 and 237.)

Greenland, S. and Thomas, D.C. 1982. On the need for the rare disease assumption in case-control studies. *American Journal of Epidemiology*, **116**, 547–553. (Cited on pp. 30 and 61.)

Greenland, S., Pearl, J., and Robins, J.M. 1999. Causal diagrams for epidemiologic research. *Epidemiology*, **10**, 37–48. (Cited on p. 7.)

Guolo, A. and Brazzale, A.R. 2008. A simulation-based comparison of techniques to correct for measurement error in matched case-control studies. *Statistics in Medicine*, **27**, 3755–3775. (Cited on p. 239.)

Gustafson, P. 2003. *Measurement Error and Misclassification in Statistics and Epidemiology*. 2nd edn. Chapman & Hall/CRC. (Cited on p. 236.)

Gustafson, P., Le, N.D., and Vallée, M. 2002. A Bayesian approach to case-control studies with errors in covariables. *Biostatistics*, **3**, 229–243. (Cited on p. 238.)

Hanley, J.A., Csizmadi, I., and Collet, J.-P. 2005. Two-stage case-control studies: precision of parameter estimates and considerations in selecting sample size. *American Journal of Epidemiology*, **162**, 1225–1234. (Cited on pp. 138 and 158.)

Hedges, L.V. and Olkin, I. 1985. *Statistical Methods for Meta-Analyses*. San Diego, Calif.: Academic Press. (Cited on p. 250.)

Heid, I.M., Küchenhoff, H., Miles, J., Kreienbrock, L., and Wichmann, H.E. 2004. Two dimensions of measurement error: classical and Berkson error in residential radon exposure assessment. *Journal of Exposure Analysis and Environmental Epidemiology*, **14**, 365–377. (Cited on p. 237.)

Hennessy, S., Bilker, W.B., Berlin, J.A., and Strom, B.L. 1999. Factors influencing the optimal case-to-control ratio in matched case-control studies. *American Journal of Epidemiology*, **149**, 195–197. (Cited on p. 82.)

Hernán, M.A., Hernandez-Diaz, S., and Robins, J.M. 2004. A structural approach to selection bias. *Epidemiology*, **15**, 615–625. (Cited on pp. 7 and 30.)

Higgins, J.P. 2008. Commentary: Heterogeneity in meta-analysis should be expected and appropriately quantified. *International Journal of Epidemiology*, **37**, 1158–1160. (Cited on p. 251.)

Higgins, J.P. and Thompson, S.G. 2002. Quantifying heterogeneity in a meta-analysis. *Statistics in Medicine*, **21**, 1539–1558. (Cited on p. 251.)

Higgins, J.P., Thompson, S.G., and Spiegelhalter, D.J. 2009. A re-evaluation of random-effects meta-analysis. *Journal of the Royal Statistical Society A*, **172**, 137–159. (Cited on p. 251.)

Hirji, K.F., Mehta, C.R., and Patel, N.R. 1988. Exact inference for matched case-control studies. *Biometrics*, **44**, 803–814. (Cited on p. 82.)

Hsieh, D.A., Manski, C.F., and McFadden, D. 1985. Estimation of response probabilities from augmented retrospective observations. *Journal of the American Statistical Association*, **80**, 651–662. (Cited on pp. 109, 110 and 157.)

Huberman, M. and Langholz, B. 1999. Re: 'Combined analysis of matched and unmatched case-control studies: comparison of risk estimates from different studies'. *American Journal of Epidemiology*, **150**, 219–220. (Cited on p. 251.)

Hughes, M.D. 1993. Regression dilution in the proportional hazards model. *Biometrics*, **49**, 1056–1066. (Cited on p. 238.)

Hui, S.L. and Walter, S.D. 1980. Estimating the error rates of diagnostic tests. *Biometrics*, **36**, 167–171. (Cited on p. 236.)

Jackson, M. 2009. Disadvantaged through discrimination? The role of employers in social stratification. *British Journal of Sociology*, **60**, 669–692. (Cited on p. 79.)

Johnston, W.T., Vial, F., Gettinby, G., Bourne, F.J., Clifton-Hadley, R.S., Cox, D.R. *et al.* 2011. Herd-level risk factors of bovine tuberculosis in England and Wales after the 2001 foot-and-mouth disease epidemic. *International Journal of Infectious Diseases*, **15** (12), e833–e840. (Cited on p. 28.)

Kalbfleisch, J.D. and Lawless, J.F. 1988. Likelihood analysis of multi-state models for disease incidence and mortality. *Statistics in Medicine*, **7**, 149–160. (Cited on pp. 188 and 210.)

Keogh, R.H. and White, I.R. 2013. Using full cohort data in nested case-control and case-cohort studies by multiple imputation. *Statistics in Medicine*, **32**, 4021–4043. (Cited on pp. 185, 186, 190 and 211.)

Keogh, R.H., Strawbridge, A., and White, I.R. 2012a. Effects of exposure measurement error on the shape of exposure–disease associations. *Epidemiologic Methods*, **1**, 13–32. (Cited on p. 238.)

Keogh, R.H., Park, J.Y., White, I.R., Lenjes, M.A.H., McTaggart, A., Bhaniani, A. *et al.* 2012b. Estimating the alcohol–breast cancer association: a comparison of diet diaries, FFQs and combined measurements. *European Journal of Epidemiology*, **27**, 547–549. (Cited on p. 243.)

Khoury, M.J. and Flanders, W.D. 1996. Nontraditional epidemiologic approaches in the analysis of gene–environment interaction: case-control studies with no controls! *American Journal of Epidemiology*, **144**, 207–213. (Cited on p. 131.)

King, G. and Zeng, L. 2001a. Explaining rare events in international relations. *International Organization*, **55**, 693–715. (Cited on p. 30.)

King, G. and Zeng, L. 2001b. Improving forecasts of state failure. *World Politics*, **53**, 623–658. (Cited on p. 28.)

King, G. and Zeng, L. 2001c. Logistic regression in rare events data. *Political Analysis*, **9**, 137–163. (Cited on p. 30.)

King, G. and Zeng, L. 2002. Estimating risk and rate levels, ratios and differences in case-control studies. *Statistics in Medicine*, **21**, 1409–1427. (Cited on pp. 60 and 109.)

Kipnis, V., Midthune, D., Freedman, L., Bingham, S., Schatzkin, A., Subar, A. *et al.* 2001. Empirical evidence of correlated biases in dietary assessment instruments and its implications. *American Journal of Epidemiology*, **153**, 394–403. (Cited on p. 237.)

Kosinski, A. and Flanders, W. 1999. Evaluating the exposure and disease relationship with adjustment for different types of exposure misclassification: a regression approach. *Statistics in Medicine*, **18**, 2795–2808. (Cited on p. 236.)

Kuha, J. 2006. Corrections for exposure measurement error in logistic regresion models with an application to nutritional data. *Statistics in Medicine*, **13**, 1135–1148. (Cited on p. 238.)

Kuha, J. and Skinner, C. 1997. *Survey Measurement and Process Quality*. New York: Wiley. Chap. Categorical data analysis and misclassification, pp. 633–670. (Cited on p. 236.)

Kuha, J., Skinner, C., and Palmgren, J. 1998. Misclassification error, in *Encyclopedia of Biostatistics*, pp. 2615–2621. 4th edn. New York: Wiley. (Cited on p. 236.)

Kulich, M. and Lin, D.Y. 2004. Improving the efficiency of relative-risk estimation in case-cohort studies. *Journal of the American Statistical Association*, **99**, 832–844. (Cited on p. 211.)

Kupper, L.L., McMichael, A.J., and Spirtas, R. 1975. A hybrid epidemiologic study design useful in estimating relative risk. *Journal of the American Statistical Association*, **70**, 524–528. (Cited on p. 209.)

Kupper, L.L., Karon, J.M., Kleinbaum, D.G, Morgenstern, H., and Lewis, D.K. 1981. Matching in epidemiologic studies: validity and efficiency considerations. *Biometrics*, **149**, 271–291. (Cited on p. 82.)

Kuritz, S.J. and Landis, J.R. 1987. Attributable risk ratio estimation from matched-pairs case-control data. *American Journal of Epidemiology*, **125**, 324–328. (Cited on p. 110.)

Kuritz, S.J. and Landis, J.R. 1988a. Attributable risk estimation from matched case-control data. *Biometrics*, **44**, 355–367. (Cited on p. 110.)

Kuritz, S.J. and Landis, J.R. 1988b. Summary attributable risk estimation from unmatched case-control data. *Statistics in Medicine*, **7**, 507–517. (Cited on p. 109.)

Langholz, B. 2010. Case-control studies = odds ratios: blame the retrospective model. *Epidemiology*, **21**, 10–12. (Cited on p. 30.)

Langholz, B. and Borgan, Ø. 1995. Counter-matching: a stratified nested case-control sampling method. *Biometrika*, **82**, 69–79. (Cited on p. 187.)

Langholz, B. and Borgan, Ø. 1997. Estimation of absolute risk from nested case-control data. *Biometrics*, **53**, 767–774. (Cited on p. 189.)

Langholz, B. and Clayton, D. 1994. Sampling strategies in nested case-control studies. *Environmental Health Perspectives*, **102** (Suppl. 8), 47–51. (Cited on p. 187.)

Langholz, B. and Goldstein, L. 1996. Risk set sampling in epidemiological cohort studies. *Statistical Science*, **11**, 35–53. (Cited on pp. 158, 187 and 188.)

Langholz, B. and Goldstein, L. 2001. Conditional logistic analysis of case-control studies with complex sampling. *Biostatistics*, **2**, 63–84. (Cited on pp. 158 and 188.)

Langholz, B. and Thomas, D.C. 1990. Nested case-control and case-cohort methods of sampling from a cohort: a critical comparison. *American Journal of Epidemiology*, **131**, 169–176. (Cited on p. 211.)

Langholz, B. and Thomas, D.C. 1991. Efficiency of cohort sampling designs: some surprising results. *Biometrics*, **47**, 1563–1571. (Cited on p. 189.)

Le Cessie, S., Nagelkerke, N., Rosendaal, F.R., van Stralen, K.J., Pomp, E.R., and van. Houwelingen, H.C. 2008. Combining matched and unmatched control groups in case-control studies. *American Journal of Epidemiology*, **168**, 1204–1210. (Cited on p. 131.)

Lee, H.J., Scott, A.J., and Wild, C.J. 2010. Efficient estimation in multi-phase case-control studies. *Biometrika*, **97**, 361–374. (Cited on p. 158.)

Levin, B. 1988. Polychotomous logistic regression methods for matched case-control studies with multiple case or control groups (Letter). *American Journal of Epidemiology*, **128**, 445–446. (Cited on p. 131.)

Liang, K.Y. and Stewart, W.F. 1987. Polychotomous logistic regression methods for matched case-control studies with multiple case or control groups. *American Journal of Epidemiology*, **125**, 720–730. (Cited on p. 131.)

Liddell, F.D.K., McDonald, J.C., Thomas, D.C., and Cunliffe, S.V. 1977. Methods of cohort analysis: appraisal by application to asbestos mining. *Journal of the Royal Statistical Society A*, **140**, 469–491. (Cited on pp. 164, 186 and 187.)

Lilienfeld, A.M. and Lilienfeld, D.E. 1979. A century of case-control studies: progress? *Journal of Chronic Diseases*, **32**, 5–13. (Cited on p. 30.)

Lin, D.Y. and Ying, Z. 1993. Cox regression with incomplete covariate measurements. *Journal of the American Statistical Association*, **88**, 1341–1349. (Cited on p. 210.)

Liu, I.-M. and Agresti, A. 1996. Mantel–Haenszel-type inference for cumulative log odds ratios. *Biometrics*, **52**, 1222–1234. (Cited on p. 131.)

Liu, M., Lu, W., and Tseng, C.-H. 2010. Cox regression in nested case-control studies with auxiliary covariates. *Biometrics*, **66**, 374–381. (Cited on p. 190.)

Lubin, J.H. and Gail, M.H. 1984. Biased selection of controls for case-control analyses of cohort studies. *Biometrics*, **40**, 63–75. (Cited on p. 188.)

Lynn, H.S. and McCullogh, C.E. 1992. When does it pay to break the matches for analysis of a matched-pairs design? *Biometrics*, **48**, 397–409. (Cited on pp. 82 and 109.)

Maclure, M. 1991. The case-crossover design: a method for studying transient effects on the risk of acute events. *American Journal of Epidemiology*, **133**, 144–153. (Cited on p. 131.)

Manski, C.F. and McFadden, D.L. 1981. *Structural Analysis of Discrete Data and Econometric Applications*. Cambridge, Mass.: MIT Press. (Cited on p. 30.)

Mansournia, M.A., Hernán, M.A. and Greenland, S. 2013. Matched designs and causal diagrams. *International Journal of Epidemiology*, **42**, 860–869. (Cited on p. 7.)

Mantel, N. 1966. Models for complex contingency tables and polychotomous dosage response curves. *Biometrics*, **22**, 83–95. (Cited on p. 130.)

Mantel, N. and Haenszel, W. 1959. Statistical aspects of the analysis of data from retrospective studies of disease. *Journal of the National Cancer Institute*, **22**, 719–748. (Cited on pp. 30 and 61.)

Mantel, N. and Hauck, W.W. 1986. Alternative estimators of the common odds ratio for 2 × 2 tables. *Biometrics*, **42**, 199–202. (Cited on p. 61.)

Maraganore, D. 2005. Blood is thicker than water: the strengths of family-based case-control studies. *Neurology*, **64**, 408–409. (Cited on p. 158.)

Marsh, J.L., Hutton, J.L., and Binks, K. 2002. Removal of radiation dose-response effects: an example of over-matching. *British Medical Journal*, **325**, 327–330. (Cited on p. 74.)

Marshall, R.J. and Chisholm, E.M. 1985. Hypothesis testing in the polychotomous logistic model with an application to detecting gastrointestinal cancer. *Statistics in Medicine*, **4**, 337–344. (Cited on p. 130.)

Marti, H. and Chavance, M. 2011. Multiple imputation analysis of case-cohort studies. *Statistics in Medicine*, **30**, 1595–1607. (Cited on p. 211.)

Martin, D.O. and Austin, H. 2000. An exact method for meta-analysis of case-control and follow-up studies. *Epidemiology*, **11**, 255–260. (Cited on p. 251.)

Mathew, T. and Nordstrom, K. 1999. On the equivalence of meta-analysis using literature and using individual patient data. *Biometrics*, **55**, 1221–1223. (Cited on p. 251.)

Mathew, T. and Nordstrom, K. 2010. Comparison of one-step and two-step meta-analysis models using individual patient data. *Biometrical Journal*, **52**, 271–287. (Cited on p. 251.)

McCullagh, P. 1980. Regression models for ordinal data (with discussion). *Journal of the Royal Statistical Society B*, **42**, 109–142. (Cited on p. 131.)

McNemar, Q. 1947. Note on the sampling error of the differences between correlated proportions or percentages. *Psychometrika*, **12**, 153–157. (Cited on p. 82.)

McShane, L., Midthune, D.N., Dorgan, J.F., Freedman, L.S., and Carroll, R.J. 2001. Covariate measurement error adjustment for matched case-control studies. *Biometrics*, **57**, 62–73. (Cited on pp. 238 and 239.)

Miettinen, O.S. 1969. Individual matching with multiple controls in the case of all-or-none responses. *Biometrics*, **25**, 339–355. (Cited on p. 82.)

Miettinen, O.S. 1970. Matching and design efficiency in retrospective studies. *American Journal of Epidemiology*, **91**, 111–118. (Cited on p. 82.)

Miettinen, O.S. 1976. Estimability and estimation in case-referent studies. *American Journal of Epidemiology*, **103**, 226–235. (Cited on pp. 30, 61 and 209.)

Miettinen, O.S. 1982. Design options in epidemiologic research: an update. *Scandinavian Journal of Work, Environment and Health*, **8** (Suppl. 1), 7–14. (Cited on p. 209.)

Miettinen, O.S. 1985. The 'case-control' study: valid selection of subjects. *Journal of Chronic Diseases*, **38**, 543–548. (Cited on p. 30.)

Mittleman, M.A., Maclure, M., and Robins, J.M. 1995. Control sampling strategies for case-crossover studies: an assessment of relative efficiency. *American Journal of Epidemiology*, **142**, 91–98. (Cited on p. 131.)

Moreno, V., Martin, M.L., Bosch, F.X., de Sanjose, S., Torres, F., and Munoz, N. 1996. Combined analysis of matched and unmatched case-control studies: comparison of risk estimates from different studies. *American Journal of Epidemiology*, **143**, 293–300. (Cited on p. 251.)

Morrissey, M.J. and Spiegelman, D. 1999. Matrix methods for estimating odds ratios with misclassified exposure data: extensions and comparisons. *Biometrics*, **55**, 338–344. (Cited on p. 236.)

Muirhead, C.R. and Darby, S.C. 1987. Modelling the relative and absolute risks of radiation-induced cancers. *Journal of the Royal Statistical Society A*, **150**, 83–118. (Cited on p. 188.)

Mukherjee, B., Liu, I., and Sinha, S. 2007. Analysis of matched case-control data with multiple ordered disease states: possible choices and comparisons. *Statistics in Medicine*, **26**, 3240–3257. (Cited on p. 131.)

Mukherjee, B., Ahn, J., Liu, I., Rathouz, P.J., and Sánchez, B.N. 2008. Fitting stratified proportional odds models by amalgamating conditional likelihoods. *Statistics in Medicine*, **27**, 4950–4971. (Cited on p. 131.)

Müller, P. and Roeder, K. 1997. A Bayesian semiparametric model for case-control studies with errors in variables. *Biometrika*, **84**, 523–537. (Cited on p. 238.)

Nan, B. 2004. Efficient estimation for case-cohort studies. *The Canadian Journal of Statistics*, **32**, 403–419. (Cited on p. 211.)

Neuhaus, J.M. and Segal, M.R. 1993. Design effects for binary regression models fitted to dependent data. *Statistics in Medicine*, **12**, 1259–1268. (Cited on p. 109.)

Neuhaus, J.M., Scott, A.J., and Wild, C.J. 2002. The analysis of retrospective family studies. *Biometrika*, **89**, 23–37. (Cited on p. 159.)

Neuhaus, J.M., Scott, A.J., and Wild, C.J. 2006. Family-specific approaches to the analysis of case-control family data. *Biometrics*, **62**, 488–494. (Cited on p. 159.)

Neuhäuser, M. and Becher, H. 1997. Improved odds ratio estimation by post-hoc stratification of case-control data. *Statistics in Medicine*, **16**, 993–1004. (Cited on p. 109.)

Neyman, J. and Scott, E.L. 1948. Consistent estimates based on partially consistent observations. *Econometrica*, **16**, 1–32. (Cited on p. 82.)

Nurminen, M. 1989. Analysis of epidemiologic case-base studies for binary data. *Statistics in Medicine*, **8**, 1241–1254. (Cited on p. 209.)

Oakes, D. 1981. Survival times: aspects of partial likelihood. *International Statistical Review*, **49**, 235–252. (Cited on p. 187.)

Olkin, I. and Sampson, A. 1998. Comparison of meta-analysis versus analysis of variance of individual patient data. *Biometrics*, **54**, 317–322. (Cited on p. 251.)

Onland-Moret, N.C., van der A, D.L., van der Schouw, Y.T., Buschers, W., Elias, S.G., van Gils, C.H. *et al.* 2007. Analysis of case-cohort data: a comparison of different methods. *Journal of Clinical Epidemiology*, **60**, 350–355. (Cited on p. 210.)

Pai, J.K., Pischon, T., Ma, J., Manson, J.E., Hankinson, S.E., Joshipura, K. *et al.* 2004. Inflammatory markers and the risk of coronary heart disease in men and women. *New England Journal of Medicine*, **16**, 2599–2610. (Cited on p. 170.)

Paneth, N., Susser, E., and Susser, M. 2002a. Origins and early development of the case-control study: part 1, early evolution. *Sozial- und Praventivmedizin*, **47**, 282–288. (Cited on p. 30.)

Paneth, N., Susser, E., and Susser, M. 2002b. Origins and early development of the case-control study: part 2, the case-control study from Lane-Claypon to 1950. *Sozial- und Praventivmedizin*, **47**, 282–288. (Cited on p. 30.)

Pearce, N. 1993. What does the odds ratio estimate in a case-control study? *International Journal of Epidemiology*, **22**, 1189–1192. (Cited on p. 30.)

Phillips, A. and Holland, P.W. 1986. Estimators of the variance of the Mantel–Haenszel log-odds-ratio estimate. *Biometrics*, **43**, 425–431. (Cited on p. 61.)

Piegorsch, W.W., Weinberg, C.R., and Taylor, J.A. 1994. Non-hierarchical logistic models and case-only designs for assessing susceptibility in population-based case-control studies. *Statistics in Medicine*, **13**, 153–162. (Cited on p. 131.)

Pike, M.C., Hill, A.P., and Smith, P.G. 1980. Bias and efficiency in logistic analyses of stratified case-control studies. *International Journal of Epidemiology*, **9**, 89–95. (Cited on p. 82.)

Prentice, R.L. 1976. Use of the logistic model in retrospective studies. *Biometrics*, **32**, 599–606. (Cited on p. 108.)

Prentice, R.L. 1982. Covariate measurement errors and parameter estimation in a failure time regression model. *Biometrika*, **69**, 331–342. (Cited on pp. 190, 211 and 238.)

Prentice, R.L. 1986a. A case-cohort design for epidemiologic cohort studies and disease prevention trials. *Biometrika*, **73**, 1–11. (Cited on pp. 209 and 210.)

Prentice, R.L. 1986b. On the design of synthetic case-control studies. *Biometrics*, **42**, 301–310. (Cited on p. 187.)

Prentice, R.L. and Breslow, N.E. 1978. Retrospective studies and failure time models. *Biometrika*, **65**, 153–158. (Cited on pp. 82 and 187.)

Prentice, R.L. and Pyke, R. 1979. Logistic disease incidence models and case-control studies. *Biometrika*, **66**, 403–411. (Cited on pp. 30, 108, 109 and 130.)

Prescott, G.J. and Garthwaite, P.H. 2005. Bayesian analysis of misclassified binary data from a matched case-control study with a validation sub-study. *Statistics in Medicine*, **24**, 379–401. (Cited on p. 236.)

Raynor, W.J. and Kupper, L.L. 1981. Category-matching of continuous variables in case-control studies. *Biometrics*, **37**, 811–817. (Cited on p. 82.)

Recchi, E. 1999. Politics as occupational choice: youth self selection for party careers in Italy. *European Sociological Review*, **15**, 107–124. (Cited on p. 27.)

Redelmeier, D.A. and Tibshirani, R.J. 1997. Association between cellular-telephone calls and motor vehicle collisions. *New England Journal of Medicine*, **336**, 453–458. (Cited on p. 125.)

Reeves, G.K., Cox, D.R., Darby, S.C., and Whitley, E. 1998. Some aspects of measurement error in explanatory variables for continuous and binary regression models. *Statistics in Medicine*, **17**, 2157–2177. (Cited on pp. 224, 225, 236, 237, 238 and 239.)

Reid, N. and Crépeau, H. 1985. Influence functions for proportional hazards regression. *Biometrika*, **72**, 1–9. (Cited on p. 210.)

Reilly, M., Torrång, A., and Klint, Å. 2005. Re-use of case-control data for analysis of new outcome variables. *Statistics in Medicine*, **24**, 4009–4019. (Cited on p. 189.)

Rice, K. 2003. Full-likelihood approaches to misclassification of a binary exposure in matched case-control studies. *Statistics in Medicine*, **22**, 3177–3194. (Cited on p. 237.)

Rice, K. and Holmans, P. 2003. Allowing for genotyping error in analysis of unmatched case-control studies. *Annals of Human Genetics*, **67**, 165–174. (Cited on p. 237.)

Ridout, M. 1989. Summarizing the results of fitting generalized linear models from designed experiments, in *Statistical Modelling: Proceedings of GLIM 89*, pp. 262–269. New York: Springer. (Cited on p. 61.)

Risch, H.A. and Tibshirani, R.J. 1988. Re: Polychotomous logistic regression methods for matched case-control studies with multiple case or control groups (Letter). *American Journal of Epidemiology*, **128**, 446–448. (Cited on p. 131.)

Robins, J.M., Breslow, N., and Greenland, S. 1986. Estimators of the Mantel–Haenszel variance consistent in both sparse data and large-strata limiting models. *Biometrics*, **42**, 311–323. (Cited on p. 61.)

Robins, J.M., Gail, M.H. and Lubin, J.H. 1986. More on 'Biased selection of controls for case-control analyses of cohort studies'. *Biometrics*, **42**, 293–299. (Cited on pp. 187 and 188.)

Robins, J.M., Prentice, R.L., and Blevins, D. 1989. Designs for synthetic case-control studies in open cohorts. *Biometrics*, **45**, 1103–1116. (Cited on p. 187.)

Robins, J.M., Rotnitzky, A., and Zhao, L.P. 1994. Estimation of regression coefficients when some regressors are not always observed. *Journal of the American Statistical Association*, **89**, 846–866. (Cited on p. 190.)

Robinson, L.D. and Jewell, N.P. 1991. Some surprising results about covariate adjustment in logistic regression models. *International Statistical Review*, **59**, 227–240. (Cited on p. 61.)

Rodrigues, L. and Kirkwood, B. 1990. Case-control designs in the study of common diseases: updates on the demise of the rare disease assumption and the choice of sampling scheme for controls. *International Journal of Epidemiology*, **19**, 205–213. (Cited on pp. 30 and 61.)

Roeder, K., Carroll, R.J., and Lindsay, B.G. 1996. A semi-parametric mixture approach to case-control studies with errors in covariables. *Journal of the American Statistical Association*, **91**, 722–732. (Cited on p. 239.)

Rosner, B., Willett, W.C., and Spiegelman, D. 1989. Correction of logistic regression relative risk estimates and confidence intervals for systematic within-person measurement error. *Statistics in Medicine*, **8**, 1051–1069. (Cited on p. 238.)

Rosner, B., Spiegelman, D., and Willett, W.C. 1990. Correction of logistic regression relative risk estimates and confidence intervals for measurement error: the case of multiple covariates measured with error. *American Journal of Epidemiology*, **132**, 734–745. (Cited on p. 238.)

Rosner, B., Willett, W. C., and Spiegelman, D. 1992. Correction of logistic regression relative risk estimates and confidence intervals for random within-person measurement error. *American Journal of Epidemiology*, **136**, 1400–1413. (Cited on p. 238.)

Rothman, K.J., Greenland, S., and Lash, T.L. 2008. *Modern Epidemiology*. 3rd edn. Philadelphia: Lippincott Williams & Wilkins. (Cited on pp. 30 and 209.)

Rubin, D.B. 1987. *Multiple Imputation for Nonresponse in Surveys*. New York: Wiley. (Cited on p. 189.)

Saarela, O., Kulathinal, S., Arjas, E., and Läärä, E. 2008. Nested case-control data utilized for multiple outcomes: a likelihood approach and alternatives. *Statistics in Medicine*, **27**, 5991–6008. (Cited on p. 189.)

Salim, A., Hultman, C., Sparén, P., and Reilly, M. 2009. Combining data from 2 nested case-control studies of overlapping cohorts to improve efficiency. *Biostatistics*, **10**, 70–79. (Cited on p. 189.)

Samanic, C.M., De Roos, A.J., Stewart, P.A., Rajaraman, P., Waters, M.A., and Inskip, P.D. 2008. Occupational exposure to pesticides and risk of adult brain tumors. *American Journal of Epidemiology*, **167**, 976–985. (Cited on p. 29.)

Samuelsen, S.O. 1997. A pseudolikelihood approach to analysis of nested case-control studies. *Biometrika*, **84**, 379–394. (Cited on p. 188.)

Samuelsen, S.O., Anestad, H., and Skrondal, A. 2007. Stratified case-cohort analysis of general cohort sampling designs. *Scandinavian Journal of Statistics*, **34** (1), 103–119, 64–81. (Cited on p. 210.)

Sato, T. 1992. Maximum likelihood estimation of the risk ratio in case-cohort studies. *Biometrics*, **48**, 1215–1221. (Cited on p. 209.)

Sato, T. 1994. Risk ratio estimation in case-cohort studies. *Environmental Health Perspectives*, **102** (Suppl. 8), 53–56. (Cited on p. 209.)

Saunders, C.L. and Barrett, J.H. 2004. Flexible matching in case-control studies of gene–environment interactions. *American Journal of Epidemiology*. (Cited on p. 158.)

Schaid, D.J. 1999. Case–parents design for gene–environment interaction. *Genetic Epidemiology*, **16**, 261–273. (Cited on p. 158.)

Scheike, T.H. and Juul, A. 2004. Maximum likelihood estimation for Cox's regression model under nested case-control sampling. *Biostatistics*, **5**, 193–206. (Cited on p. 190.)

Schill, W. and Drescher, K. 1997. Logistic analysis of studies with two-stage sampling: a comparison of four approaches. *Statistics in Medicine*, **16**, 117–132. (Cited on pp. 157 and 158.)

Schill, W., Jöckel, K.H., Drescher, K., and Timm, J. 1993. Logistic analysis in case-control studies under validation sampling. *Biometrika*, **80**, 339–352. (Cited on p. 239.)

Scott, A.J. and Wild, C.J. 1986. Fitting logistic models under case-control or choice-based sampling. *Journal of the Royal Statistical Society B*, **48**, 170–182. (Cited on pp. 109, 157 and 158.)

Scott, A.J. and Wild, C.J. 1991. Logistic regression models in stratified case-control studies. *Biometrics*, **47**, 497–510. (Cited on pp. 131 and 157.)

Scott, A.J. and Wild, C.J. 1997. Fitting regression models to case-control data by maximum likelihood. *Biometrika*, **84**, 57–71. (Cited on pp. 131 and 157.)

Scott, A.J. and Wild, C.J. 2001. Case-control studies with complex sampling. *Journal of the Royal Statistical Society C*, **50**, 389–401. (Cited on p. 157.)

Scott, A.J. and Wild, C.J. 2002. On the robustness of weighted methods for fitting models to case-control data. *Journal of the Royal Statistical Society B*, **64**, 207–219. (Cited on p. 158.)

Seaman, S.R. and Richardson, S. 2004. Equivalence of prospective and retrospective models in the Bayesian analysis of case-control studies. *Biometrika*, **91**, 15–25. (Cited on p. 109.)

Self, S.G. and Prentice, R.L. 1988. Asymptotic distribution theory and efficiency results for case-cohort studies. *Annals of Statistics*, **16**, 64–81. (Cited on p. 210.)

Siegel, D.G. and Greenhouse, S.W. 1973. Validity in estimating relative risk in case-control studies. *Journal of Chronic Diseases*, **26**, 210–225. (Cited on p. 109.)

Siegmund, K.D. and Langholz, B. 2001. Stratified case sampling and the use of family controls. *Genetic Epidemiology*, **20**, 316–327. (Cited on p. 159.)

Smith-Warner, S.A., Spiegelman, D., Ritz, J., Albanes, D., Beeson, W.L., Bernstein, L. *et al.* 2006. Methods for pooling results of epidemiologic studies: the pooling project of prospective studies of diet and cancer. *American Journal of Epidemiology*, **163**, 1053–1064. (Cited on p. 251.)

Solomon, P.J. 1984. Effect of misspecification of regression models in the analysis of survival data. (Amendment **73** (1986), 245). *Biometrika*, **71**, 291–298. (Cited on p. 189.)

Spiegelman, D., McDermott A., and Rosner, B. 1997. Regression calibration method for correcting measurement-error bias in nutritional epidemiology. *American Journal of Clinical Nutrition*, **65**, 1179S–1186S. (Cited on p. 238.)

Steenland, K. and Deddens, J.A. 1997. Increased precision using countermatching in nested case-control studies. *Epidemiology*, **8**, 238–242. (Cited on p. 187.)

Støer, N. and Samuelsen, S. 2012. Comparison of estimators in nested case-control studies with multiple outcomes. *Lifetime Data Analysis*, **18**, 261. (Cited on p. 189.)

Stuart, E. 2010. Matching methods for causal inference. *Statistical Science*, **25**, 1–21. (Cited on p. 82.)

Stukel, T.A., Demidenko, E., Dykes, J., and Karagas, M.R. 2001. Two-stage methods for the analysis of pooled data. *Statistics in Medicine*, **20**, 2115–2130. (Cited on p. 251.)

Sturmer, T. and Brenner, H. 2002. Flexible matching strategies to increase power and efficiency to detect and estimate gene–environment interactions in case-control studies. *American Journal of Epidemiology*. (Cited on p. 158.)

Sutton, A.J. and Higgins, J.P.T. 2008. Recent developments in meta-analysis. *Statistics in Medicine*, **27**, 625–650. (Cited on p. 251.)

Therneau, T.M. and Li, H. 1998. Technical Report Series No. 62, Computing the Cox model for case-cohort designs. Section of Biostatistics, Mayo Clinic, Rochester, Minnesota. (Cited on p. 210.)

Thomas, D.C. 1977. Addendum to: Methods of cohort analysis: appraisal by application to asbestos mining. *Journal of the Royal Statistical Society A*, **140**, 469–491. (Cited on p. 187.)

Thomas, D.C and Greenland, S. 1983. The relative efficiencies of matched and independent sample designs for case-control studies. *Journal of Chronic Diseases*, **10**, 685–697. (Cited on p. 82.)

Thomas, D.C., Goldberg, M., Dewar, R., *et al.* 1986. Statistical methods for relating several exposure factors to several diseases in case-heterogeneity studies. *Statistics in Medicine*, **5**, 49–60. (Cited on p. 130.)

Thomas, L., Stefanski, L., and Davidian, M. 2011. A moment-adjusted imputation method for measurement error models. *Biometrics*, **67**, 1461–1470. (Cited on p. 238.)

Thompson, S.G. and Sharp, S.J. 1999. Explaining heterogeneity in meta-analysis: a comparison of methods. *Statistics in Medicine*, **18**, 2693–2708. (Cited on p. 251.)

Thompson, S.K. 1992. *Sampling*. New York: Wiley. (Cited on p. 261.)

Thompson, W.D., Kelsey, J.L., and Walter, S.D. 1982. Cost and efficiency in the choice of matched and unmatched case-control designs. *American Journal of Epidemiology*, **116**, 840–851. (Cited on p. 82.)

Thürigen, D., Spiegelman, D., Blettner, M., Heuer, C., and Brenner, H. 2000. Measurement error correction using validation data: a review of methods and their applicability in case-control studies. *Statistical Methods in Medical Research*, **9**, 447–474. (Cited on p. 237.)

Ury, H.K. 1975. Efficiency of case-control studies with multiple controls per case: continuous or dichotomous data. *Biometrics*, **31**, 643–649. (Cited on p. 82.)

VanderWeele, T.J., Hernández-Diaz, S., and Hernán, M.A. 2010. Case-only gene–environment interaction studies: when does association imply mechanistic interaction? *Genetic Epidemiology*, **34**, 327–334. (Cited on p. 131.)

Wacholder, S. and Hartge, P. 1998. Case-control study, in *Encyclopedia of Biostatistics, Volume 1*, pp. 503–514. Chichester: Wiley. (Cited on pp. 29 and 30.)

Wacholder, S., Gail, M.H., Pee, D., and Brookmeyer, R. 1989. Alternative variance and efficiency calculations for the case-cohort design. *Biometrika*, **76**, 117–123. (Cited on p. 210.)

Wacholder, S., McLaughlin, J.K., Silverman, D.T., and Mandel, J.S. 1992a. Selection of controls in case-control studies, I. Principles. *American Journal of Epidemiology*, **135**, 1019–1028. (Cited on p. 30.)

Wacholder, S., McLaughlin, J.K., Silverman, D.T., and Mandel, J.S. 1992b. Selection of controls in case-control studies, III. Design options. *American Journal of Epidemiology*, **135**, 1042–1051. (Cited on p. 82.)

Wang, M. and Hanfelt, J.J. 2009. A robust method for finely stratified familial studies with proband-based sampling. *Biostatistics*, **10**, 364–373. (Cited on p. 159.)

Weinberg, C.R. and Sandler, D.P. 1991. Randomized recruitment in case-control studies. *American Journal of Epidemiology*, **134**, 421–432. (Cited on p. 108.)

Weinberg, C.R. and Umbach, D.M. 2000. Choosing a retrospective design to assess joint genetic and environmental contributions to risk. *American Journal of Epidemiology*, **152**, 197–203. (Cited on p. 158.)

Weinberg, C.R. and Wacholder, S. 1993. Prospective analysis of case-control data under general multiplicative intercept risk models. *Biometrika*, **80**, 461–465. (Cited on p. 109.)

Whitaker, H.J., Farrington, C.P., Spiessens, B. and Musonda, P. 2006. Tutorial in biostatistics: the self-controlled case series method. *Statistics in Medicine*, **25**, 1768–1797. (Cited on p. 131.)

Whitaker, H.J., Hocine, M.N., and Farrington, C.P. 2009. The methodology of self-controlled case series studies. *Statistical Methods in Medical Research*, **18**, 7–26. (Cited on pp. 128 and 131.)

White, I.R. and Royston, P. 2009. Imputing missing covariate values for the Cox model. *Statistics in Medicine*, **28**, 1982–1998. (Cited on p. 190.)

White, I.R., Royston, P., and Wood, A.M. 2011. Multiple imputation using chained equations: issues and guidance for practice. *Statistics in Medicine*, **30**, 377–399. (Cited on p. 189.)

White, J.E. 1982. A two stage design for the study of the relationship between a rare exposure and a rare disease. *American Journal of Epidemiology*, **115**, 119–128. (Cited on p. 157.)

Whittemore, A.S. 1982. Statistical methods for estimating attributable risk from retrospective data. *Statistics in Medicine*, **1**, 229–243. (Cited on p. 110.)

Whittemore, A.S. 1995. Logistic regression of family data from case-control studies. *Biometrika*, **82**, 57–67. (Cited on p. 158.)

Whittemore, A.S. and Halpern, J. 1989. Testing odds-ratio equality for several diseases. *Statistics in Medicine*, **76**, 795–798. (Cited on pp. 130 and 131.)

Whittemore, A.S. and Halpern, J. 1997. Multi-stage sampling in genetic epidemiology. *Statistics in Medicine*, **16**, 153–167. (Cited on pp. 156 and 158.)

Whittemore, A.S. and Halpern, J. 2003. Logistic regression of family data from retrospective study designs. *Genetic Epidemiology*, **25**, 177–189. (Cited on pp. 158 and 159.)

Wild, C.J. 1991. Fitting prospective regression models to case-control data. *Biometrika*, **78**, 705–717. (Cited on pp. 131, 157 and 158.)

Willett, W. 1998. *Nutritional Epidemiology*. 2nd edn. Oxford University Press. (Cited on p. 237.)

Witte, J.S., Gauderman, W.J., and Thomas, D.C. 1999. Asymptotic bias and efficiency in case-control studies of candidate genes and gene–environment interactions: basic family designs. *American Journal of Epidemiology*, **149**, 694–705. (Cited on p. 158.)

Woolf, B. 1955. On estimating the relationship between blood group and disease. *Annals of Human Genetics*, **19**, 251–253. (Cited on p. 61.)

Wright, S. 1921. Correlation and causation. *Journal of Agricultural Research*, **20**, 162–177. (Cited on p. 7.)

Xie, Y. and Manski, C.F. 1989. The logit model and response-based samples. *Sociological Methods and Research*, **17**, 283–302. (Cited on p. 30.)

Yaghjyan, L., Colditz, G.A., Collins, L.C., Schnitt, S.J., Rosner, B., Vachon, C. *et al.* 2011. Mammographic breast density and subsequent risk of breast cancer in postmenopausal women according to tumor characteristics. *Journal of the National Cancer Institute*, **103**, 1179–1189. (Cited on p. 112.)

Yates, F. and Cochran, W.G. 1938. The analysis of groups of experiments. *Journal of Agricultural Science*, **28**, 556–580. (Cited on p. 250.)

Zhang, J. and Borgan, Ø. 1999. Aalen's linear model for sampled risk set data: a large sample study. *Lifetime Data Analysis*, **5**, 351–369. (Cited on p. 189.)

Zhang, L., Mukherjee, B., Ghosh, M., Gruber, S., and Moreno, V. 2008. Accounting for error due to misclassification of exposures in case-control studies of gene–environment interaction. *Statistics in Medicine*, **27**, 2756–2783. (Cited on p. 237.)

Zhao, L.P., Hsu, L., Holte, S., Chen, Y., Quiaoit, F., and Prentice, R.L. 1998. Combined association and aggregation analysis of data from case-control family studies. *Biometrika*, **85**, 299–315. (Cited on p. 159.)

Zhou, H. and Pepe, M.S. 1995. Auxiliary covariate data in failure time regression. *Biometrika*, **82**, 139–149. (Cited on pp. 190 and 211.)

Zota, A.R., Aschengrau, A., Rudel, R.A., and Brody, J.G. 2010. Self-reported chemicals exposure, beliefs about disease causation, and risk of breast cancer in the Cape Cod Breast Cancer and Environment Study: a case-control study. *Environmental Health*, **9**, doi:10.1186/1476–069X–9–40. (Cited on p. 213.)

Index